Series / Number 02-036

International Networks: The Structured Webs of Diplomacy and Trade

BONNIE H. ERICKSON
University of Toronto

SAGE PUBLICATIONS / Beverly Hills / London

Printed in the United States of America

For information address:

SAGE PUBLICATIONS, INC.
275 South Beverly Drive
Beverly Hills, California 90212

SAGE PUBLICATIONS LTD
St George's House / 44 Hatton Garden
London EC1N 8ER

International Standard Book Number 0-8039-0526-2

Library of Congress Catalog Card No. 75-9034

FIRST PRINTING

When citing a professional paper, please use the proper form. Remember to cite the correct Sage Professional Paper series title and include the paper number. One of the two following formats can be adapted (depending on the style manual used):

(1) AZAR, E. E. (1972) "International Events Interaction Analysis." Sage Professional Papers in International Studies, 1, 02-001. Beverly Hills and London: Sage Pubns.

OR

(2) Azar, Edward E. 1972. *International Events Interaction Analysis.* Sage Professional Papers in International Studies, vol. 1, series no. 02-001. Beverly Hills and London: Sage Publications.

CONTENTS

BONNIE H. ERICKSON, Assistant Professor of Sociology at the University of Toronto, received her Ph.D. from Harvard University. She has written several articles for the Canadian Review of Sociology and Anthropology *and is currently writing a statistics textbook,* Understanding Data, *with T. A. Nosanchuk of Carleton University. Dr. Erickson is completing projects on the belief systems of Canada and the structure of secrecy networks.*

International Networks: The Structured Webs of Diplomacy and Trade

BONNIE H. ERICKSON
University of Toronto

I. SYSTEMIC HYPOTHESES IN INTERNATIONAL STUDIES

To expand understanding of international systems, this paper attempts to develop and test a related pair of novel systemic hypotheses. An international system is understood to be a set of bounded and interdependent relationships between nations. Trade relationships in 1947-1948 form one such system, with trade between any two nations being partly dependent on trade among other nations, and diplomatic links in the same period form another system in which links affect each other; these are the two data sets for which systemic hypotheses are tested here. Both the hypotheses predict the size of a link (whether trade dollar volume or number of diplomats exchanged) in terms of other related links in the system. One hypothesis predicts that a link between two nations will be large to the extent that the two nations involved have similar patterns of linkage to other nations. The other hypothesis predicts that the link size is related to the overall size of links among other nations relatively close to the two whose link is predicted. The bounded and interdependent nature of a system is reflected in these hypotheses, for they spell out one possible kind of link interdependence and they attempt to explain system

AUTHOR'S NOTE: *This paper and the dissertation on which it is based would not have been possible without a great deal of help from many people and institutions. The Canada Council and Carleton University provided financial support. Bill Bradley and Randall Levitt of the Carleton University Computer Center are responsible for the difficult programming. Many helpful criticisms and suggestions were provided by Hayward Alker, Lynn Mytelka, T. A. Nosanchuk, Phil Stone, Donald P. Warwick, and, above all, Harrison C. White. Three anonymous Sage reviewers made excellent suggestions for publication revision.*

components (relationships) in terms of other system components rather than in terms of external factors such as individual attributes of nations.

There is very wide agreement that international relationships are systemic in some way, and many systemic studies have been done, but the particular kind of systemic hypothesis used here fills a gap in the previous array of studies.

A few examples will suffice to illustrate the many cases in which a scholar argues for the existence of an international system but does not develop the idea in the same way as the present study. Haas (1964: 7) suggests:

> The brooding omnipresence of some species of system simply cannot be banished. What then do we mean by system? Following in the footsteps of Nadel . . . I conceive of a system as a second level body of abstractions, a network of relationships among relationships, leaving the concrete one step below.

However, Haas is more concerned with explaining international organizations and their goals than in explaining international relationships or establishing their systemic character:

> The system consists of the relationships among the *patterns* of trade, war, migration, and subversion as they impinge upon one another in the confines of international organizations [Haas, 1964: 53].

Kaplan (1961) notes that much-discussed structures, such as balance of power, may only be discussable as abstracted patterns of relations among relations; patterns of alliance formation, as opposed to alliances in themselves. Yet Kaplan's main work on international systems (1957), including balance of power systems, relies on a normative approach in which different systems are derived from different "rules of the game." Singer (1963) notes that dyadic studies of one sort of international relation, power, are less useful than a more complex relational view, since:

> The international system is neither a dyad (duopoly) nor a multiplicity of dyads . . . all are influencing all, directly or indirectly [Singer, 1963: 421].

But Singer uses a dyadic model anyway because it is more manageable.

Some studies do focus on the explanation of international relationships, like this one, but they tend to be either much more global or much more atomistic in approach. On the one hand are the global studies which work out the overall structure of some set or sets of links. For example Russett (1967) examines world regions, stressing their stability over different kinds of relationships and over different time periods. Galtung (1966) stresses the prediction of individual links to a greater degree but uses global patterns to do so; a link depends on where two nations fit into a world system of status and cliques, with the nature of the dependence being a function of the overall nature of the system (feudal for example). Studies like these model the world as a whole at the cost of underplaying local variations. On the other hand, there are studies which account for relationships atomistically, without any systemic component at all. For example, Linneman (1966) relates trade volumes to individual nation traits like gross national product. Studies like these can be seen as rivals to the systemic ones, for they attempt to explain the same relationships without making use of any systemic variables. Thus atomistic approaches can be used as sources of alternate or benchmark hypotheses.

Between the global and the atomistic lies the intermediate territory to be opened up here. The rest of this chapter will describe the two systemic hypotheses and their mechanism in general terms, while detailed operationalization of the crucial variables will be postponed until the next chapter. Now a reasonable first step in formulating systemic hypotheses is to make some sensible but simple assumption about how links affect other links. Some such guiding assumption is essential, since there are endless ways in which links could be related to each other.

As a simple and general rationale for the hypotheses to be developed, assume that a link in a network (of trade or of diplomacy) is a potential channel of communication between the two nations involved; this channel is especially likely to transmit information related to the link content. Thus two nations which trade a good deal will probably exchange a lot of information about trade; two nations which exchange many diplomats will probably exchange a lot of information about diplomatic affairs. Information passed on in one link can be further passed on in another, going from nation to nation in chains of communication.

These chains provide a form of indirect contact between nations which will affect their direct contact. If two nations are involved in many strong chains of indirect communication, they will tend to have a stronger direct link; the indirect contacts will help provide knowledge about interaction opportunities and will encourage similarities between the two which make interaction easier. On the other hand, if two nations are only weakly linked indirectly then they will tend to have a weak direct link as well because of the lack of support and pressure from their part of the international network. This view of the importance of indirect communication cannot be directly tested here, but I can point out that many students of international relations have taken similar positions. For example Russett (1967: 124) argues that "trade becomes a channel of communication . . . manifested in responsiveness."

Supposing that the general assumption is workable, it still has to be converted into specific hypotheses relating indirect contacts to direct links. At this point it is very helpful to turn to the interpersonal literature, which contains some extensive work on two very promising hypotheses of the appropriate sort. These prior efforts have saved a good deal of possible effort for this study by offering a good start on conceptualization and by suggesting some important difficulties and possible solutions. Further, the usefulness of the major hypotheses in both areas–in both interpersonal and international relations–adds a good deal to their value by showing their generality of application and breadth of empirical support. Thus I have no hesitation in siding with those who urge application of suitable sociological hypotheses to international material (such as Galtung, 1968) rather than those who argue that the international arena is an entirely distinct problem (like Hoffman, 1965). Some international events may be distinctive enough to resist application of general hypotheses, perhaps because the events depend on the views and goals of a few mutable individuals, but most international interaction is more routine and predictable. As McClelland (1966: 104) argues:

> A dense network of transactions at the base of the international system is made up of administered and routinized activities and is substantially self-organizing and self-directing.

The final decision on such strategy debates has to be pragmatic, so it is worthwhile to note that this study is not the first to find interpersonal hypotheses very useful in tackling international topics. Alker and Russett (1965) use the concept of cross-pressure to help build an explanation of voting in the United Nations General Assembly; Singer (1963) draws hypotheses from small group studies; Galtung (1964) uses status inconsistency as a predictor of aggression; and Singer (1974) has recently been trying for very great generality by drawing on general systems theory for possible explanations of war.

All in all, it seems that more vigorous exploitation of sociological sources should be fruitful for understanding international systems and for developing more general, more widely supported hypotheses. Two sociological research areas which focus on interpersonal networks are especially useful here: one tradition stemming from the work of Elizabeth Bott (1971) on marriage and social networks, and another tradition stressing friendship patterns. Bott's work provides the basic conceptualizations to be used here while the friendship literature offers some useful supplementary material, and both sources offer suggestions as to how the systemic effects might work. Thus the next step is to describe the hypotheses derived from these sources and to consider which of the suggested mechanisms for them seem most applicable to international systems.

Where the dependent variable of this study is the strength of links between nations, Bott's dependent variable was the strength of links between spouses. At one extreme, a husband and wife have a segregated role relationship if they share few tasks or leisure activities; at the other extreme they have a joint relationship if they share most of their activities. Since the amount of interaction (rather than, say, love) is what we are interested in, the segregated-joint distinction boils down to a weak-strong continuum of interaction for present purposes.

Bott argued that the strength of marital interaction depended on the pattern of links in the husband's and wife's network of acquaintances, much as a nation-nation link is related to its location in an international system here. Bott's major accomplishment was the invention of two flexible and powerful concepts capturing important aspects of the surrounding relationships: the concepts of *overlap* and *knitness.* For any pair in a system of relationships, *overlap* is the extent to which the pair members are related to the same other members of the system. In particular, the overlap in a couple's set of acquaintances is the proportion of acquaintances who are known by both the husband and the wife. The greater the overlap, the closer the marital relationship. This result parallels friendship studies; for example, it is well known that people tend to be closer if they have friends in common.[1] Now, why might overlap be associated with a pair's level of interaction? Higher overlap, or more shared friends, means more channels of indirect communication are available. The wife talks to a friend, who talks to the husband; or one friend talks to another who passes ideas on to a third who is linked to the first. High levels of such indirect interaction foster higher levels of shared norms, of similar tastes, and of positive feelings.

Bott's other independent variable was *knitness,* "the extent to which the people known by a family know and meet one another independently of the family." More generally, any pair in a system of relationships will have links to other members of the system. (These links may involve one or both members of the pair; overlap may be low or high.) The proportion of possible links among these other members which actually exist is knitness. At first Bott thought that high knitness went with segregated marriage, but further work indicated that this result needed clarification.[2] In Bott's cases high knitness went with low overlap; that is to say, the husband's set of friends was densely tied together and so was the wife's, but the two sets had few links between them. This situation is parallel to that for two people who belong to different friendship cliques and who are thus not likely to interact a great deal. Both cases are easily understood in terms of indirect communication: the pair members interact in different circles, becoming similar to and attracted to those in their different circles; they have few if any intercircle channels to connect them indirectly and pull them together. Things are different when the pair members are exposed to high knitness among other system participants linked to both the pair members, for example, when two people belong to the same tightly knit group of friends. Then indirect fostering of the pair's relationship is considerable.

Thus the interpersonal literature provides a clear warning: knitness must be refined. Let us distinguish internetwork knitness (the extent to which two actors' sets of friends have ties which link those two sets) from intranetwork knitness (the extent to which each actor's friends have ties within the actor's set). All these links are part of Bott's original idea of knitness, since all are links between the actors' friends and do not involve the actors themselves; yet clearly they may have very different implications. Links between actors' cliques pull them together, since they are to some degree in the same clique; links that tend to be confined within cliques pull the actors apart.

Overall, then, family and friendship studies suggest three relational hypotheses. For any pair in a system of relationships:

Hypothesis 1—The higher the overlap, the more the pair interacts.

Hypothesis 2—The higher the internetwork knitness, the more the pair interacts.

Hypothesis 3—The higher the intranetwork knitness, the less the pair interacts.

These hypotheses have considerable support at the interpersonal level, they are general enough to be extended to the international levels, and the discussion above has sketched a rationale general enough to work at both levels.

Before going further, a more detailed discussion of possible rationales is necessary—partly to see whether the rationale has implications for the research strategy, partly to make sure that transferring the hypotheses between such disparate areas is reasonable, and partly to point to related ideas which should be tested at some point even though they are beyond the scope of this study.

The selection of a possible mechanism for the hypotheses is made easier by the fact that some of the interpersonal level suggestions are clearly inappropriate for an international system. For example, Bott at some point talks of a "lump of sociability"–a limited amount of interaction capacity–as an explanation for the effects of knitness on conjugal ties. If a spouse is absorbed in an intensive set of interactions outside the family, he may just not have much left over for interaction with the other spouse. But this reasoning is less applicable to organizations or nations which can increase their interactive capacities by, say, recruiting more diplomats or revaluing currency to encourage trade. Granted, nations cannot increase these capacities indefinitely, especially if they are relatively weak, but they are in a far better position to do so than individuals.[3]

In contrast the idea of indirect communication provides a far more general rationale. The idea is meaningful for either the interpersonal or the international level. Further, all three hypotheses can be related to this notion since overlap and both forms of knitness can be seen as summary measures of indirect communication paths (a point which will be further clarified in the third chapter as the operational measures of these variables are discussed). So far, communication has been discussed in very general terms. By drawing in a more specific model of a possible mechanism for communication effects, some specific implications can emerge, useful in this study and possible related ones.

A mechanism which is more specific, yet flexible enough to cover interpersonal and international cases, can be found in Homans' (1950) general theory of interaction and Simon's (1952) mathematical rephrasing of this theory. Put very simply, interaction in a group is postulated to have direct or indirect reciprocal relationships with friendliness and activity. Simon refers to the latter two variables as "causes of communication." The group level of activity, and hence of friendliness and interaction, is also affected by the external environment through pressures toward greater activity. This model can be applied to the present problem by assuming that the "group" is the pair whose interaction is to be predicted, and the "external pressures" are just what overlap and knitness sum up: the extent to which the pair is surrounded by intensive interaction, which generates high levels of activity and friendship and communication, which encourage interaction between the pair members.

One immediate payoff of the Simon model of interaction is that this model clarifies the role of non-links, of cases in which two nations do not trade or exchange any diplomats. As Simon points out, it takes more pressure to start an interaction than to maintain it at an appropriate level, once begun. Thus, nations which do not interact, which have no recorded trade of diplomatic exchange, cannot be expected to fit the hypotheses. This simplifies the present study's job considerably by reducing the number of links to be explained. Other studies have set aside noninteracting pairs, often without an explicit theoretical reason. For example, Linneman (1966) left pairs without reported trade out of his econometric analysis, but did so because of measurement difficulties. Brams (1966), following Goodman (1963) also left out such pairs but without explicit reasons. In this study, the contrast between pairs with zero or nonzero links is traced to the hypothesis that starting an interaction may take more pressure than is needed to maintain it. At some point not easily specified, the pressure should take effect, an interaction should begin, and once begun it should settle down at a level appropriate to the pressures represented by the level of overlap and knitness. It follows that the first noninteracting pairs to begin relationships should be those with the highest overlap and knitness. Thus there are two indirect ways of testing the appropriateness of the mechanism suggested here: one could ask whether zero pairs behave differently from nonzero ones, and whether the zero pairs with high knitness and overlap are the first to start nonzero ties.

The behavior of zero pairs was partly explored here in a sample of nation pairs more fully discussed in the operationalizations chapter. To foreshadow the sample's results briefly, zero pairs did behave very differently from nonzero ones: zero pairs showed no sign of relationship to knitness and overlap (which ranged widely for these pairs), while nonzero pairs showed expected relationships to the relational variables.

However, it was beyond the scope of this study to pursue the other mechanism test suggested—using overlap and knitness to predict which zero pairs would start interaction first. Predictions of this sort would be worth following up since they offer ways to link international behavior from year to year without falling back on sheer persistence (such as an assumption that the basic regional structure of the world tends to stay the same) and political accident (for example, such-and-such a regional modification emerged because of a war or an international upheaval or whatever). We can predict, on a reasonable basis, where the network of international interaction will begin to fill in. We can also make predictions about future behavior of pairs which do interact in a given time period: the level of interaction is often somewhat above or below the predicted value and there should be some pressure on that level to get closer to the prediction. To put this another way, Simon's formulation describes a group in equilibrium—and we know very well that international systems are in partial equilibrium at most. Thus we can look for pairs that bend the equilibrium expectations (pairs with high overlap and no interaction for example) and predict that they will tend to move toward those expectations. At this point I cannot claim that such predictions would work, and they would be computationally expensive–but they do suggest nontrivial diachronic possibilities reflecting both the systemic character of international relationships and their imperfect degree of equilibrium.

The Simon-Homans model offers a few other implications for future testing. Here, only levels of interaction (trade and diplomacy) are examined. What of the other elements of the model? One might treat friendliness as a predisposition reflected in policy statements or generated by similarities (like common culture). Activity, as distinguished from trade, diplomacy, or other forms of interaction, might be indexed by joint action in military pacts or development programs. External pressure toward activity could have another interpretation as political pressure, such as a great power persuading its dependents to form political or economic organizations. In short, it is possible to operationalize the elements of the Simon-Homans model with data available in long time series, which would allow a more direct test of this potential mechanism for the three systemic hypotheses.

An alternative line of development would be to work out other possible rationales, perhaps rationales that were themselves systemic. For example, one might consider how communication from nation to nation takes place. Probably this would require study of the relevant substructures within nations (where nations are "black boxes" here) and would suggest differences from one interaction system to another because different relations are handled by different subsystems; (here the hypotheses are applied to trade and diplomacy in the same ways.) The international--interpersonal parallel would break down, and so would the assumption that international systems are closed as well as interdependent.

There is one more general point to raise here: how can one tell if "it works"–if the hypotheses are well founded? The data to be used are complete interaction matrices for the world's nations, a population rather than a sample, so that inferential statistical tests are of no use or meaning.[4] One cannot simply compare results with what one would expect from chance fluctuations in samples, since we have a census rather than a sample. But one can compare results to those expected from an alternative model. Here, and indeed in any systemic study using complete data, this is the only strategy that will give a plausible test of predictions.

In particular, the systemic hypotheses should be compared to non-systemic ones in a competitive way to see whether the systemic ones have anything to add. The main point here is one of scientific parsimony: systemic hypotheses are more complex than others so there is little point in retaining them unless they make some contribution to justify the extra effort. Too often a complex systemic model is compared only to a random or null model (if any explicit comparison is made); this is not a very powerful test of a model's accuracy. We do not have to decide between a patterned international system and random noise, but between two possible sources of patterning: the interdependence of nations and the atomistic characteristics of nations. To the extent that systemic hypotheses do things which nonsystemic hypotheses cannot do, we get strong support for the idea of "international system."

In the long run the systemic and nonsystemic factors influencing international relationships can be merged. The previous step only tries to establish a distinct systemic element, not to show that the system explains everything: international systems are partly open rather than fully closed. Note that this step and the previous one may be merged in practice. For example, one might have a model of the combined systemic and nonsystemic factors showing that both make strong contributions. Such a model would show how the two factors merge, while also showing that the systemic factor could not be ignored. But the two steps should be separated in thinking things through even if merged in an actual study. To construct an alternative model, one assumes that a pair's relationship is a function of the individual properties of the pair members. There are many such properties on which one might rely. Both international and interpersonal studies have investigated the effects of status, cultural similarity, proximity, and so forth, on interaction. Bales (1951) reports on a property which is strongly tied to interpersonal interaction and which works well for complete interaction matrices. Bales shows that pairs of group members interact in proportion to their combined overall interaction activity. If both pair members interact a great deal, then they are likely to interact highly with each other. If both interact little, then they are likely to interact little with each other. If one interacts heavily but the other not, then they are likely to interact to a moderate degree with each other and so on. More exactly, consider any pair of nations. The total interaction for the first nation (say its total foreign trade) is multiplied by the total interaction for the second nation (again, say its total foreign trade) and the product is divided by the overall total interaction (say the world's total trade). This is a parsimonious, reasonable, and easily calculated prediction and it makes no use of the relational properties which are to be tested. It is also a fairly powerful representative of possible nonsystemic approaches. For example, Linneman (1966) uses a model which involves the gross national products of a pair of nations in a way that parallels this study's use of interaction totals (that is, total international trade). In addition, he uses six other variables, some of which (like trade preference areas) are not dyadic. In predicting international trade, his model achieves multiple correlations of about 0.80, while the far simpler model used here has zero order correlations of comparable size. Even noting that Linneman distinguishes direction of trade, so that trade from A to B is separated from trade from B to A, and also noting that he uses data from a different time period (1958-1960, averaged), the comparison is still quite favorable to the simpler model used here.

Conclusions

This chapter has developed systemic hypotheses, a general rationale and mechanism for them, and a nonsystemic alternative to serve as a benchmark for the usefulness of the more complex systemic approach. Further, all of these elements have been designed so that they are meaningful for either interpersonal or international systems. Work on international systems benefits because the interpersonal literature offers useful concepts and models which have already been partly worked out and debugged—a considerable advantage in approaching anything as complex as a study of relationships. As long as the parallels seem to hold, still more can be drawn from the interpersonal literature.[5] On the other hand, the interpersonal tradition stands to gain just as much, since the lively ideas generated there have not been matched with adequate data. Collecting detailed, accurate, complete information on an interpersonal network is a formidable job, rarely undertaken, and still more rarely achieved.[6] But on the international level, virtually complete and very detailed records of trade, diplomacy, and other networks are available.

The very richness of the data poses a problem in transferring the three relational hypotheses from people to nations. Trade and diplomacy are quantitatively measured, where Bott (like most interpersonal researchers) had only dichotomies to work with. Her measures are simple to work out when dealing either with people who know each other or not. More complex measures are needed for nations which interact to a certain degree or not; the strength of a link, not just whether there is one, has to be worked in. A more complex reworking of overlap and knitness will follow the next chapter, which describes the data.

II. THE DATA

The hypotheses and underlying mechanism spelled out in the previous chapter were stated in very general terms, with no limiting conditions specified. *Any* interaction network could be used to test these predictions: there is no need to stick with the interpersonal relationships from which the hypotheses were derived. Bott makes much the same point herself in discussing her family study:

> To me the fact that it talks about families and networks is more or less incidental. The basic problem it deals with is the way a group, any sort of group, is related to its environment. My aim was (and still is) to understand how the internal functioning of a group is affected not only by its relationship *with* the people and organizations of its environment, but also by the relationships among those people and organizations [Bott, 1971: 249].

Thus, the difficult part of finding a set of data to test her propositions is not to find an appropriate set, but to find a feasible set, one which can be collected in a reasonable time. The problem is that testing the effects of overlap and knitness requires full information about the relationships among all the people known by the pairs of people whose ties one would like to predict. Even for a small number of pairs, this is a very large number of relationships. Probably the best source of a large number of pairs, plus full data on their networks, is a moderately large group with reasonably clear-cut boundaries. If reasonably large, there will be plenty of pairs within it; if well bounded, the pair networks will not expand outward indefinitely, but be confined within the group itself.

Thus international relations data are suitable for the ideas worked out for interpersonal relations; the international data also have the convenient properties which interpersonal data often lack. The set of countries of the world forms a clearly delimited group; both participants in and students of the world scene tend to insist that international relations are more or less sui generis and can be analyzed apart from intranation relationships. Further, the international group is of a usable size: about one or two hundred nations since World War II. At the time dealt with in this study, there were about one hundred countries in the world, a number small enough to work with but large enough to guarantee more than enough pairs for analysis. Data are not only fully available, with records of almost all interactions in the group, but they are available for more than one kind of interaction—so the generality of the predictions can be more fully explored. Nearly complete and quite accurate records were available for two types of international interaction: trade and diplomacy.

Trade data are available from United Nations reports for many years, but data on diplomatic exchanges are available for only two years in detailed form. Analysis of both years with full diplomatic data was beyond the scope of this study, so the question became which of the two to use. Data for 1963-1964 were collected by Alger and Brams (1967), but were not yet available from the Inter-University Consortium for Political

Research. In any case, use of the earlier 1948-1949 data allows predictions to be made about which noninteracting pairs of the earlier period will most likely be interacting pairs in the later period, which has a much lower amount of noninteraction. Predictions of this sort will eventually help test the mechanism behind the hypothesis. Data on 1948-1949 diplomacy were drawn from *The Diplomatic Yearbook* (United Nations World, 1951), which has nearly-complete lists of diplomatic staff throughout the world at that time. These data have been coded as Alger and Brams suggest, that is, the link between two nations is measured by the total number of diplomats exchanged between them. Only diplomats resident in the capital of the country to which they are accredited are counted, since non-residents are not likely to be very effective. If the number of diplomats sent by country A to country B is known, but the number sent from B to A is not, the latter is estimated by the former.

Trade data are taken from the *United Nations Statistical Papers Series T, Direction of World Trade, Vol. X.* Trade is reported in both directions by both partners in most cases; here the average of the two reports of the export plus import totals is used, or the available report is used if one country does not report. In a few cases, neither country reported. Data from 1949 were used instead to fill in the gaps for all of South Korea's trade and for Israeli and Ethiopian trade with other non-reporting countries of 1948. No such estimates were available for trade links among the U.S.S.R., Albania, Rumania, Saudi Arabia, and Thailand, so these links were treated as missing data. Links involving "China" were omitted from the analysis because it was rarely clear whether trade figures under this heading referred to China or Taiwan.

It should be noted that others have made a somewhat different coding decision, basing total trade on the sum of exports only, on the grounds that import data often include transportation costs (see Russett, 1967: 128). This is somewhat inconsistent, since import data must be used in some cases when a non-reporting country does not provide the required export data, and it does not allow for the variety of factors that may be involved in the difference between B's report of its imports from A, and A's report of its exports to B. A may have a different currency conversion schedule, a more or less inclusive definition of reportable trade, and so on; perhaps the most important factor is the variable quality of national data collection and reporting. On the whole, it seems best to use all the data available. The difference between this coding procedure and Russett's should not be great.

The data, especially the trade data, are very highly skewed: there are few very high values and any number of small ones. Indeed, nearly half the pairs of countries have zero trade or diplomatic links. To make the very skewed trade distribution more manageable, trade link sizes were transformed into logarithms (two-decimal logarithms with the decimal point dropped). This should have rather little effect on the analysis which will use the relative sizes of links more often than the absolute sizes.

Students of international relations since World War II will no doubt point out that 1948-1949 was by no means a routine year in world history. A major war was just completed, the losers of it were still largely excluded from normal international relations, many nations were still struggling back from the economic devastation of wartime, and so on. Results shown for this year may not be fully general. But on the other hand, results shown for a year so little likely to be fully "self-organizing" may well be very strong results indeed. It is not easy to say just how the oddities of the post-war period are related to the conditions which might make or break the hypotheses to be tested, whether the 1948-1949 conditions favor the hypotheses, hinder them, or are merely irrelevant. This is not the fault of the historically-oriented, who can discuss the peculiar features of this time, but of the theorizers who have not spelled out the conditions of the systemic model more fully. Given this lack, one can only go on with testing the hypotheses on the year available—with plans to continue the testing further on different sorts of years and different sorts of international systems—in the hope of specifying the conditions of the hypotheses.

International specialists may also point out that the coding of links used here makes an important and controversial assumption about the nature of links, an assumption that the asymmetric aspect of links can be safely ignored. Now most links are obviously asymmetric; a great power sends far more diplomats and far more export trade to a little power than the little power returns. Some, including Russett (1967), argue that one must take this into account in the very way that links are measured. Russett likes to index the relative impact of a link on nations of different sizes by norming the link on some measure of the nation's capacity. For example, a trade flow might be divided by the nation's domestic gross product. If the two nations involved in a given trade flow have different domestic economic size, then the link will be different in importance for each of them and will be measured differently too. This presents some difficulties in analysis, such as separate derivations and discussions of "chooser" and "chosen" matrices (looking at these asymmetrically-measured links first in one direction, and then in another).

The practical difficulties of using asymmetric links in the present study would be considerable; for a start, there would be twice as many link values to cope with. But more important is the fact that there is nothing in the theory used here to suggest that links must be treated as asymmetric, nor anything to suggest how asymmetric links should be analyzed. Rather, the stress on the role of communication implies that asymmetry can in fact be ignored. It matters little that the United States sends far more diplomats to small Latin American countries than the small countries return, since all of these diplomats act as communicators in *both* directions. Trade flows between a pair of nations may be unbalanced, yet communication be even, since a flow in either direction offers chances of information exchange to both parties.

The assumption that a trade or other interaction link has variable effect on countries of variable size is parallel to Bott's "lump of sociability" hypothesis. The crucial assumption is that country's attention is limited by its domestic market size or the size of its foreign service; this limited amount must be shared out among the various interaction links competing for attention. [6A] But it is also possible to argue that complex organizations can alter their attention capacity, or that smaller nations may actually be less affected by a link because their internal communication channels are less sophisticated. The point cannot be decided by argument alone. Here, links will be treated as symmetrical and will not be normalized on "capacities" except partially in passing; this decision could be wrong and will have to be checked by further research.

In conclusion, the trade and diplomatic data fulfill the requirements of the hypotheses and rationale used and, on the other hand, the hypotheses should help make sense of a little-known aspect of the data: the "local structure" of routine international interaction. The next step is to bring the hypotheses and data together by developing new operationalizations geared to quantitative relationships. The next chapter describes and illustrates such operationalizations, briefly compares them to alternative possibilities, and provides a partial check on the Simon-Homans rationale which motivates some of the basic data analytic decisions.

III. SPECIFYING THE SYSTEMIC VARIABLES

Adapting the simple forms of overlap and knitness used by Bott is made much simpler by the notion of indirect channels of communication. It is possible to construct new operationalizations which use the quantitative measures of interaction, and the full networks of interaction, which are available for trade and diplomacy. Although these new forms are more complex than the ones developed for mere 0-1 measures of relationships, they are equivalent to those original forms when only simple measures of relationships are available.

The explanation of the overlap and knitness operationalizations will probably be clearer with an example, so consider the data in Table 1. Using these figures, overlap and internetwork knitness and intranetwork knitness will be calculated for the USA and Canada. Note that these examples are deliberately much simpler than, and different from, the systemic calculations that will be used later for the USA and Canada; for example, the later calculations will be based on the world-wide trade system, not just Central and North America.

Overlap sums up the potential strength of the shortest possible paths of indirect communication, paths involving only one third party per path. In our example, the shortest indirect paths between the USA and Canada are:

USA – Costa Rica – Canada

USA – Cuba – Canada

USA – Dominican Republic – Canada

The indirect paths will be strong possible channels of communication when they are composed of strong links. If the USA has a strong trade relationship with Costa Rica, and so does Canada, then communications can flow between the USA and Canada via Costa Rica. But if Canada has a strong tie with Costa Rica and the USA has a weak one, or vice versa, the communication will be short-circuited; it will get to Costa Rica but no further. For strong communication channels, the best arrangement of USA and Canada trade ties is one in which they both have their strong links to the same nations and their weak ones to the same nations; the worst arrangement is one in which the USA trades strongly with just those nations with which Canada is tied weakly, and vice versa. In short, the question is the extent to which the two nations have similar patterns of interaction with third parties; this is just the extent to which their trade relationships are correlated. Note that rank order rather than product moment correlations are used here, since the latter are too sensitive to extreme or erratic values, which are frequent in international data sets.[7] In our example, we want the rank order correlation between USA trade and Canadian trade, which is just the correlation between the first and second column of Table 1 (not including the 3485 value which is USA-Canada trade, or the relationship whose size is to be predicted with the systemic variables). The Spearman *r* works out to .90, and this corresponds to a visual impression that the USA and Canada have similar trade profiles in this region.

There is one other point about the computation of overlap which does not come up in this example but does matter in the data analysis proper: the correlation is only computed for those third party nations which have

TABLE 1
North and Central American Trade

	USA	CANADA	COSTA RICA	CUBA	DOMINICAN REPUBLIC	EL SALVADOR	GUATEMALA	HAITI	HONDURAS	MEXICO	NICARAGUA
USA											
CANADA	3485										
COSTA RICA	62	3.0									
CUBA	802	30.0	.3								
DOMINICAN REPUBLIC	84	20.0		.3							
EL SALVADOR	62	2.0	.3	.2							
GUATEMALA	93	7.0		.2		2.0					
HAITI	41	1.3		.1							
HONDURAS	44	4.0	.2	2.0	.2	5.0	.2				
MEXICO	809	30.0	.6	12.0	.4	.7	4.0	.2	1.0		
NICARAGUA	36	.5	.4	.5		.7	.1		.1	.8	
PANAMA	80	3.0	.3	.4		.4	.1		.1	1.0	.9

nonzero ties to one or both of the nations whose link is being predicted. For example, if neither the USA nor Canada had any recorded trade with Panama then Panama would have been ignored in finding overlap. Most of the entries in the data matrices are zero, so that inclusion of double zeros would inflate a correlation substantially. For example, two nations could attain a high correlation merely by being so small that they both interact with few others—a not uncommon situation! It makes more sense to confine the correlations to values for partners that are demonstrably partners for at least one member of the pair, that is, have nonzero interaction with one or both. When double zeros crop up, they are treated as missing data (dropped from the correlation, with appropriate reduction in N). This procedure is consistent with the earlier remarks on the difference between pairs with and without interaction. Earlier the important aspect of the difference was that noninteracting nations could not be expected to fit the hypotheses because they would not begin interaction and smooth responsiveness to external pressures until the external pressures had built up to a fairly high level. Here the use of the contrast is a little different: if neither of the pair members interacts with a given third nation, then the non-links with this nation exert no pressure on either member of the pair. That is, noninteracting nation pairs are right out of the picture: they neither respond to knitness and overlap in the predicted way nor contribute to the interaction pressures which these variables represent.

Turning now to the knitness variables, the ideal operationalizations would be directly parallel to the one for overlap: knitness would also measure the potential strength of indirect paths of communication, the only difference being that knitness would sum up longer paths than overlap. In our example, longer paths would include ones like:

Canada – Costa Rica – Cuba – USA
Canada – El Salvador – Haiti – USA

Some paths would be even longer than these, with three or more intermediaries. The path strengths could be measured in much the same way as the short overlap paths by

(1) finding these longer paths;
(2) evaluating path strength in some way, perhaps by multiplying the link sizes in each path;
(3) combining path strengths into an overall measure of communication lines with two or more intermediaries.

In principle there are many ways of doing these things but all of the ways are impractically time-consuming. In Table 1, for example, there are ten nations besides the USA and Canada so there are 10(9) = 90 paths with two intermediaries, and far more if longer paths are considered. To avoid prohibitively expensive calculations, some radical simplification is necessary.

The heart of the simplification is to look only at the crucial links in a chain, not at the whole chain. For example, one can get some idea of whether the path

Canada – Costa Rica – Cuba – USA

is likely to be strong by asking how strong the link Costa Rica – Cuba is, since any communication by this path will have to get through this link. Of course, a strong link between Costa Rica and Cuba would be of no use if the other links in the chain were weak, so only links between nations which have substantial ties to at least one of Canada and the USA need to be considered. To put this another way, it is plausible that a pair is not much influenced by interactions among nations of no importance to either pair member. Thus, attention can be limited to the five nations with which Canada trades the most and the five with which the USA trades the most, or to the ten with which Canada trades the most and the ten with which the USA trades the most, or, in general, to the "top k" interaction partners of each member of the pair.[8]

The question, of course, is which *k* to use–or where to set the cutoff point for an "effective neighborhood" of nations whose interactions are likely to matter to the pair whose link is going to be predicted. One could try knitness measures using various values of *k* and pick the *k* that gave the best results, but this would be too post hoc. Instead, one can take advantage of the fact that overlap and knitness variables are based on different links, so if the best size of *k* is found from *overlap* then there is nothing post hoc about using that *k* for *knitness.* So the general strategy is to compute overlap values for various possible effective neighborhoods and pick the neighborhood that most closely reproduces the results for the basic overlap measure.

One might limit the neighborhood to links above a certain size, such as the expected value of links with the nation in question. But the distribution of link sizes varies widely from one nation to the next, so that the number of interaction partners above such a criterion value would vary also. Let links be restricted by the less variable criterion of rank; that is, only a nation's top five or ten or *k* links will be examined. But this is not quite all, since the top *k* links for both nations in a pair have to be considered. So find the top k interaction partners for (say) Canada and the top k for (say) the USA, merge the two lists of partners, and then find overlap for the given value of k. Suppose we are working with Table 1 and $k = 3$. The top three trading partners for the USA are Mexico, Cuba, and Guatemala, while the top three for Canada are Mexico, Cuba, and the Dominican Republic. If the two lists are merged, the neighborhood overlap for $k = 3$ is found by getting the rank order correlation of these values:

	Trade with USA	Trade with Canada
Mexico	809	30
Cuba	802	30
Guatemala	93	7
Dominican Republic	84	20

The length of the merge list used in the final calculation will vary. If two nations have quite different top k partners, then the merged list will be $2k$ long; while if they have just the same partners the merged list will be k long.

In general, the overlaps for different list sizes do not have to be the same. If they vary erratically with respect to list length itself, then it will be very difficult to choose a boundary for the effective network or neighborhood, since there will be no way to guess which list length is giving the "right" answer. Rather, one might conclude that there is no right answer to give, because the phenomenon is too unstable. But if the values for list overlap change regularly as the list lengthens, then it may be possible to say that there is some reasonable breaking point. We shall see that there is a reasonable breaking in the trade and diplomatic data for $k =$ 15 or 20.

Thus the knitness measures will be based on the overall strength of relationships among those nations on a pair's top k list of interaction partners, which will give a necessarily crude indication of the probable strength of indirect paths involving two or more of these top k nations. But as has been argued already in distinguishing internetwork and intranetwork knitness, the effect of strong links in this effective neighborhood may be variable: some may fit into chains linking a pair, while some may pull the pair apart because one of the pair members is connected to these strongly linked third parties while the other is not. The difference can be more clearly exemplified from Table 1 if we use Canada and Costa Rica for the pair whose systemic variables are to be calculated: this pair has weaker overlap than the USA and Canada. Suppose we take k = 5 for this pair and get the following trade ranks:

	Trade with Canada	Trade with Costa Rica
USA	1	1
Cuba	2.5	5
Mexico	2.5	2
Nicaragua	10	3
El Salvador	8	5
Panama	6.5	5
Guatemala	4	9
Honduras	5	7

The first three nations are strongly linked to both Canada and Costa Riça, the next three only to Costa Rica, and the last two only to Canada. Strong links among or between these different groups will have different implications for possible communication chains.

First, there are intermediary links that fit the general concept internetwork knitness because, if strong, they can lead to strong indirect communication between the pair members. Certainly this includes links among nations with strong ties to both the pair members, that is, the first three nations listed above. For example, if Cuba and Mexico are strongly tied then the path

USA – Cuba – Mexico – Canada

or, for that matter,

USA – Mexico – Cuba – Canada

will be strong overall. The end links are known to be strong so the overall effect of the path just depends on the middle section.

Internetwork links also include links between the networks special to each of the pair members, or, in this case, links between Guatemala and Honduras on the one hand and Nicaragua, El Salvador, and Panama on the other. Again, if these links are strong then indirect communication can be strong. For example if Panama and Guatemala are well linked then the chain

Costa Rica – Panama – Guatemala – Canada

should be strong since the other two links in it are already known to be of respectable size.

The remaining links are all intranetwork: they are links within a set of nations tied to just one of the pair members. If such links are strong they draw a pair member into a separate subgroup rather than drawing it toward the other pair member. Thus one would expect that the Canada-Costa Rica link would be weakened, not strengthened, by strong links among Nicaragua, El Salvador, and Panama or between Guatemala and Honduras.

In general, suppose there is a pair of nations, A and B. For the desired value of *k,* the merged list of top interaction partners of A and B will have three possible types of entries: nations in both A's top *k* and B's top *k,* called C_i; plus nations on A's top list only, called A_i; plus nations on B's top list only, called B_i. *Internetwork knitness* or *KI* is the level of interaction for $C_i - C_i$ and $A_i - B_i$ links, which can be the middle sections of long indirect chains of communication. *Intranetwork knitness* is the level of interaction of $A_i - A_j$ links *(KA)* and $B_i - B_j$ links *(KB),* which are more likely to draw A and B apart than together if strong.

To complete the operational definitions of the knitness measures only a choice of link level is needed. Here, the overall strength of interaction among a set of third party nations is found by taking their trimean,[9] a robust combination of the first three quartiles. Overall, then, the operational forms of overlap and knitness sum up some crucial systemic variables: the extent to which a nation pair is likely to get pressure toward interaction through indirect communication channels, where the channels can be shorter (overlap) or longer (knitness).

For example, one can find KI and KA + KB for K = 5 for the pair (Canada-Costa Rica) used as an example above. The links on which KI is based are:

USA – Cuba	802
USA – Mexico	809
Cuba – Mexico	12
Nicaragua – Guatemala	.1
Nicaragua – Honduras	.1
El Salvador – Guatemala	2
El Salvador – Honduras	5
Panama – Guatemala	.1
Panama – Honduras	.1

Here $KI = \frac{12 + .1 + 2(2)}{4} = 4$. The links on which KA+KB is based are:

Nicaragua – El Salvador	.7
Nicaragua – Panama	.9
El Salvador – Panama	.4
Guatemala – Honduras	.2

and $KA{+}KB = \frac{.8 + .3 + 2(.55)}{4} = .55.$

In the rest of this chapter the measures developed above will be compared to some other possible forms of overlap and knitness, and they will be used to make a partial indirect check on the accuracy of the Simon-Homans rationale.

Equivalence to Measures for 0-1 Relationships

First of all, it is useful to know that the more complex operations worked out for quantitatively measured links will reduce to the traditional operations when only 0-1 measurement is available. Thus results here are directly comparable to those based on dichotomously measured relationships. This includes not only the interpersonal studies in the Bott tradition, but potential international studies by those who prefer to dichotomize international relationships. For example, Brams (1966) categorized links as 1 or 0, present or absent, according to the "significance" or "non-significance" of their size. He argued strongly that links do not matter until they get beyond a certain point (an argument used in a modified form in developing the knitness operationalizations).

If links are measured only by 0s and 1s then the present form of overlap and the traditional measure are perfectly related. The higher the proportion of third parties linked to both the pair members, the higher the rank order correlation between the two sets of links. (Third parties which are not linked to either pair member are not considered in either form of overlap.)

With 0-1 links the present form of knitness differs only slightly from the traditional form based on proportions. Traditionally the knitness in a group is the proportion of possible links which exist, or the ratio of "1" links to all links, or the mean link size. The present form of knitness uses the trimean rather than the mean and will thus give similar but more discontinuous results. For 0-1 links the traditional form of knitness may be preferable because it is more sensitive. As argued above, using mean link size with quantitative links would not be robust enough.

Some Other Approaches to Overlap and Knitness

This section will briefly report the performance of some alternate operationalizations for samples of trade and diplomacy pairs. Those who are not concerned about technical variations may wish to skim the description of the samples, which are also used in the next section, and skip over the rest of this section.

The nations were numbered and one hundred pairs of them were chosen by means of a random number table. There was one restriction on nations eligible for this draw: only those which appear in both the diplomatic and trade data were accepted, in order to facilitate comparability. This restriction biases the sample somewhat in two not very serious ways. First, sampled nations must indeed be nations (or else they will not appear in the diplomatic data). Many students of international affairs would argue that one ought to focus on nations in any case, since interactions involving dependent territories are poorly recorded. They are also ambiguous in that one could make an argument for counting those interactions as part of the ruling country's activities, at least as well as for counting them as the dependent country's. These problematic data are left in the overall arrays, since they can be quite important for some internation contacts. For example, Asian nations trade actively with Asian non-nations, especially Hong Kong.

The second bias comes from the exclusion of nations that do have recorded diplomatic interaction, but not recorded trade. These countries (Afghanistan, German Democratic Republic, North Korea, Liberia, Monaco, Nepal, San Marino, and Vietnam) were too small or too troubled (or both) to have much international interaction of any kind in the postwar years. One is not likely to lose much by excluding them.

Another sort of bias is the absence of high extreme values in the samples. There are only a few very large trade or diplomatic links, so it is perhaps not surprising that none turned up in the preliminary sample.

Since the pairs were randomly sampled from all possible pairs, aside from the exceptions noted, many of them turned out to be unrelated. Only 40 of the pairs had nonzero diplomatic ties and 69 of them had nonzero trade ties. These nonzero cases are used in the rest of this section, while both the nonzero and zero cases are useful in the next section's check on the Simon-Homans rationale.

One way to vary the operational form of overlap has already been indicated: the rank order correlation can be limited to a merged list of top k interaction partners. The rationale for such *neighborhood correlations* would be that most third parties simply do not count; the objection would be that the best k could only be found by arbitrary fiat or ex post facto. The correlations based on varying list lengths do not have to have similar values and well may not.

One gets a parallel set of options by ignoring the quantitative variations in nonzero link sizes, returning to the older stress on mere list overlap. The *total list overlap* is the number of third parties linked (with nonzero ties) to both members of a pair, as a proportion of third parties linked to either member of the pair. This measure parallels the basic correlational overlap measure. One could also look at list overlaps based only on top k lists of interaction partners, paralleling the neighborhood correlations.

Comparative results for the trade and diplomatic samples are shown in Table 2. Note that the correlations within this table are product-moment correlations indicating the performance of various overlap measures, some of which are rank order correlations. First one sees that the "total r," or the basic measure of overlap, is indeed related to interaction strength as hypothesized. There is a product-moment correlation of .40 with diplomatic link strength and .64 with trade link strength. As a harder test, consider just the residuals from the atomistic null prediction:

$$\text{Residual} = \text{Link Size} - \frac{(\text{Total Interaction for A})\ (\text{Total for B})}{\text{World Total}}$$

Even when the atomistic prediction (a powerful one) has been removed, overlap correlates .65 with diplomatic residuals and .44 with trade residuals.

The other possible overlap measures also have the predicted positive relationships with links and residuals, in all but two cases. This suggests that overlap's connection with trade and diplomacy is not just an artifact of a particular measurement choice. On the other hand, none of the rival measures are related to links or residuals with the consistent strength shown by total r, which suggests that total r is the right measure to use.

TABLE 2
Overlap Measures Correlations for Nonzero Pairs

	Correlations for Diplomatic:			Correlations for Trade:		
Measure	Links	Null Residuals	Total r	Links	Null Residuals	Total r
Total r	.40	.65		.64	.44	
r, k = 5	.16	.13	.26	.02	.25	.26
r, k = 10	.00	.23	.56	.06	.31	.29
r, k = 15	.03	.40	.72	.14	.37	.43
r, k = 20	.05	.49	.81	.20	.45	.57
Total LO	.09	.19	.34	.50	-.05	.41
LO, k = 10	.01	.41	.66	.28	.43	.45
LO, k = 15	.08	.51	.82	.22	.42	.63
LO, k = 20	.18	.53	.85	.27	.45	.65

Based on the 40 nonzero link pairs in the diplomatic sample and the 69 nonzero pairs in the trade sample.

The total list overlap measure (total LO), nearest to the traditional approach to overlap, is very erratic. It has a respectable relationship to trade (.50) but a weak one to diplomacy (.09); its correlations with trade and diplomatic residuals are only –.05 and .19. All in all, this is an inferior measure.

The neighborhood measures have a rather different pattern: it is the LO measures, not the r measures, which perform better. For a given value of *k*, the relationship between the list overlap measure and links or residuals is always a little stronger than the relationship between the neighborhood correlation measure and the same dependent variable. This result is encouraging for those who need a more economical approach and reflects well on the interpersonal studies which have used variants of list overlap perforce.

If one must use a neighborhood measure of overlap, a simple list overlap based merely on dichotomous links may well suffice. But the neighborhood measures are clearly inferior to "total r" overlap.

First, they require a difficult bootstrap decision: what *k* defines the neighborhood? There is no obvious theoretical basis for choice, but there are some pragmatic strategies. First, one could find a representative subset of the data (like these preliminary samples) and find both total correlational overlap and list overlaps for various feasible values of *k* up to the value which could actually be used in the final research. Then compare these and choose the overlap that performs best. Finally, carry through the data collection for the required value of *k;* for example, one might ask people to name their best *k* friends or closest *k* business acquaintances, or whatever. It could be a serious mistake to choose such a value without validating it first.

Another strategy would be to predict *k* on the basis of another study, but this seems a risky line to take since the behavior of overlap with list length may well vary in some unknown way with, say, the extensiveness of the networks.

The neighborhood measures are also inferior in performance: none of them, not even list overlap for higher *k* values, has as consistently strong ties to links and residuals as the "total r" measure. This is not unreasonable since the total correlational measure uses more of the relevant information than any other measure. In fact, the more of this information that a measure uses, the better it tends to do; the neighborhood measures tend to improve in performance as the list length on which they are based grows. One striking argument for the total correlation as a measure of overlap is that the other measures do better and better as they grow closer and closer to the total correlation (relationships between rival measures and "total r" are also shown in Table 2).

Thus one can come to the happy conclusion that the overlap measure most clearly related to the theory is also the best in practice. The total correlation uses the most data information, makes the most consistently strong predictions, requires no difficult decisions about arbitrary *k* values, and appears to act as the criterion for the effectiveness of the other measures. The closer an alternative measure of overlap is to the total correlation, the better a predictor of the dependent variables it will be. It is also true that the alternative measures, especially some of the list overlap measures, do often come very close to the total correlation measure. In practical terms, this implies that researchers may not have to get data on the strength of every bond in the network they are studying, but may instead be able to make do with data on the presence or absence of links in the neighborhood of links to be analyzed. Conceptually, this result that the various suggested measures of overlap are all reasonably close to each other implies that the concept of overlap can be defined in various useful and more or less interchangeable ways. Results using overlap are not likely to be artifacts of the particular method chosen for measuring the variable.

The overlap variable has resemblances to other approaches that use or could be adapted to use the same sort of data. It is similar to the one-step chain component of a mean first passage time, for both use some function of the cross products of pair members' links to third parties:

$$(A - C_i) \times (B - C_i).$$

Both are akin to the Morisette (1958) procedure for finding the strength of an unknown relation by multiplying known relations in a cycle (in the present situation the link from A to C_i and the link from B to C_i would be the two known relations from the cycle formed by relations among A, B, and C_i). In all these procedures, the cross products are normed in different ways, but are supposed to answer the same question: do A and B interact with other nations (or people, or whatever) in the same way? If so, they should interact considerably with each other, because otherwise the system is unbalanced and strain will be felt (Morisette and other balance theorists), or because their overlapping networks make them socially close (Beshers and Laumann, 1967), or because the indirect communication fosters interaction as argued here. One could also make a comparison between overlap and other ways of predicting the strength of an unknown relation from known relations in a cycle, ways using some function of the discrepancy between the known links. A rank order correlation is such a function, since its major term is the sum of the squared differences between the ranks of A's and B's links to others; a very similar expression appears in Taylor (1970: 219).

Several alternate forms of knitness were also explored (sometimes by accident, after miscommunication with the computer programmer, rather than by plan). None of the alternate forms were as satisfactory as the main versions of internetwork and intranetworks knitness, versions which are themselves only approximate, so the alternate forms will be discussed only very briefly.

One very simple alternative is to follow Bott again and use everything in sight. Take all the nations on either pair member's list of *k* top partners (the merged list used for neighborhood overlaps) and find all the links between them. Then find the trimean of these links. Since this measure is the overall level of interactions among nations in a *union* of two lists, it is called the knitness of the union or just KU. The measure was inadequate for Bott's or Turner's data (Bott, 1971; Turner, 1967) in which high values of KU (for dichotomous links) gave a misleading impression. The high knitness values came from high levels of interaction within, but not between, the separate networks of the pair members (the husband and wife). Since each spouse belonged to a distinct network, the high levels of interaction pulled them apart rather than together. Something similar happens here, where a KU measure based on quantitative links produces very weak and fluctuating correlations with links or residuals from the atomistic model, whereas one gets quite good results when knitness is partitioned into intranetwork and internetwork components, or KA and KB versus KI.

For example, the choice of effective neighborhood, or of *k*, turns out to be painless for the knitness operations preferred here. KI, the most important knitness variable, is nearly insensitive to list length. The overlap results in Table 2 suggest that a list length of at least $k = 15$ or $k = 20$ should be used, since only then do the neighborhood overlaps become substantially related to the criterion form "total r." On the other hand, one would like to use as small a *k* as possible to save calculation and to avoid defining effective neighborhoods so broadly that one pair's is much like another's. Thus it is agreeable to find that KI varies only slightly with list length so that $k = 15$ (a modest size) can be safely used. KI for $k = 15$ correlates highly with KI for $k = 20$ on the one hand (correlating at .87 for trade and .88 for diplomacy) and also correlates highly with KI for $k = 10$ on the other hand (correlating at .85 for trade and .83 for diplomacy). Other measures, like KU, lack this robustness with respect to list length.

KI also shows some modest predictive power in the preliminary samples. Using $k = 15$, KI has correlations of .216 and .132 with trade and diplomatic links. If the atomistic predictions are removed, KI actually works better; it has correlations of .445 and .446 with trade and diplomatic residuals from the atomistic model. KA and KB are of minor importance, producing small negative correlations with links and residuals whatever variation of knitness is being used.

A major possible variation can be constructed simply by calculating KI, KA, and KB in the same way except for one thing: the measures are trimeans of residuals from the atomistic model, not of raw interaction sizes as before. That is, one can ask not how great the level of third party interaction is in absolute amount, but how great it is in comparison to the other interactions in which this set of nations is involved. The measures used so far reveal only the overall level of interaction for a given set of nations. They do not indicate whether or not this level is high considering the extent to which this given set interacts in general, a consideration important for those who argue that the impact of a link depends on the size of a nation (for example, Russett, 1967). A set of nations in the merged neighborhood of a given pair may have a fairly high level of interaction and yet have little effect on that pair because the links making up that high level are minor ones from the point of view of the nations in the neighborhood. If they are minor, they are relatively unlikely to serve as channels of information and other variables that could foster interaction between the members of the focal pair.

Interesting as this view is, it did not work out very well in the preliminary samples. Results were never better than those for knitness based on raw link sizes; they were often worse. Overall they were much less consistent. For those who favor the residual-based knitness strongly for theoretical reasons, it may be useful to know that good results seem most likely when KI is restricted to relationships among nations on both A's and B's top k interaction partner lists, that is, the intersection of the top partner lists for both members of a pair. For example, suppose overlap, KI, KA, and KB are used in multiple regressions to predict trade and diplomatic residuals from the null model. If the knitness variables are based on residuals, with KI based on the intersection of the two lists, the multiple correlations are .539 for trade residuals and .657 for diplomatic residuals. Very similar multiple correlations of .526 and .668 are found when the main knitness operations based on links are used. Other comparisons between link-based and residual-based knitness are favorable to the former, however.

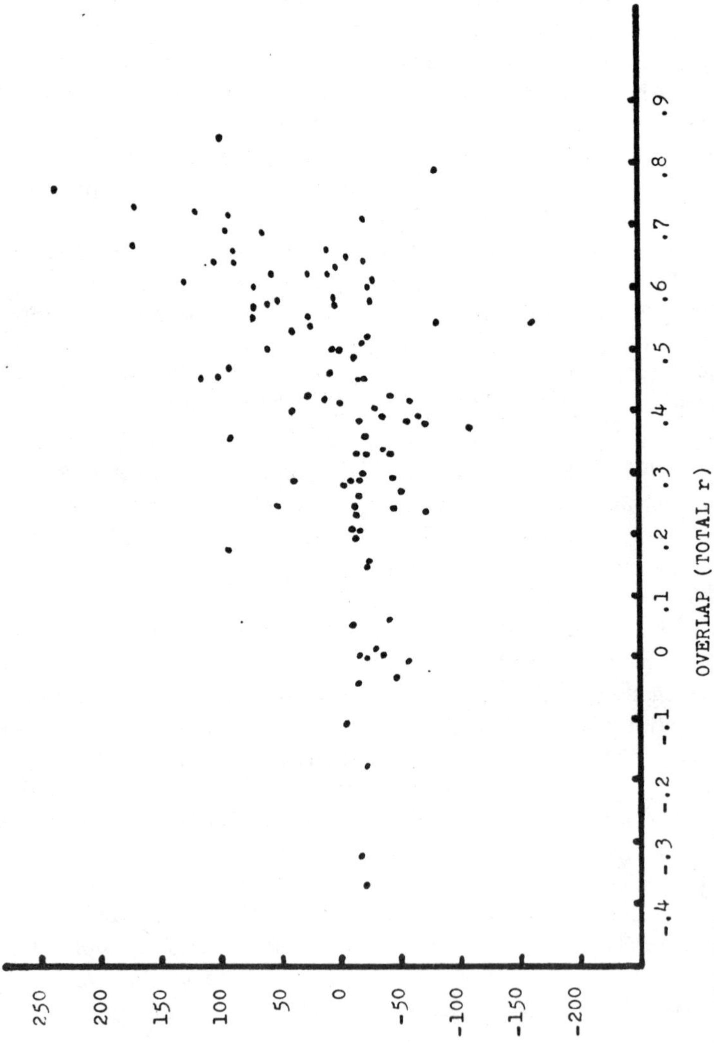

Figure 1: Trade residuals from null model, 100-pair sample

A Partial Check of a Rationale

The rationale drawn from the Simon-Homans model suggested that nation pairs with no interaction should be different from those with interaction; in particular, the former need not fit the hypotheses about overlap and knitness that are supposed to help explain the latter. The importance of this distinction can be seen dramatically in a sequence of plots: 1, 2, 3A, 3B, 4A, 4B. Figures 1 and 2 use overlap as the independent variable (it is the most powerful network variable, after all); the dependent variable is trade residuals in Figure 1 and diplomatic residuals in Figure 2. (If trade and diplomatic links are used, the following results are even more striking.) Both these figures are odd-looking for scatterplots; most plots look like watermelons or cucumbers, while Figures 1 and 2 look more like tilted Vs. The problem is that these figures are based on all the 100 pairs of the preliminary samples, so that nonzero pairs are mixed with zero pairs even though the rationale suggests they must be treated separately. The remaining figures show what happens when the separation is made. Figure 3A uses pairs with zero trade links and Figure 4A uses pairs with zero diplomatic links; both show that residuals from the atomistic model are not at all related to overlap for these noninteracting pairs. If links are used instead of residuals, the soft bars in 3A and 4A become solid lines–all the pairs have zero links although varying widely in overlap.

Figures 3B and 4B give plots for trade and diplomacy pairs which *do* interact, and here overlap is related to the residuals in standard-looking scatterplots. If 3A is smoothed over 3B, or 4A over 4B, one sees how the peculiar V-shaped plots for all pairs (Figures 1 and 2) arose from combining two visibly different classes of nation pairs. This is a striking support for the rationale used here, and is also support for those who have excluded zero pairs from analysis almost automatically.[10]

Another strategy for showing the importance of analyzing nonzero pairs is less visual than the previous strategy but more comprehensive.

The systemic and atomistic variables were used in regressions to find out how much they could account for all together. Table 3 shows the various multiple correlations which show: first, that the full set of variables accounts for a very large amount of the variance in trade or diplomacy; second, that adding in systemic variables picks up a large proportion of the variance which the atomistic prediction leaves unexplained; and third, that the systemic variables add most to the null model when only nonzero links are considered. For the nonzero pairs, the atomistic model accounts for 80% of the diplomatic variance, while the atomistic plus the systemic variables account for 88%. The atomistic accounts for 58% of the trade variance, while the atomistic plus the systemic variables account for 73%. This is as it should be. The atomistic

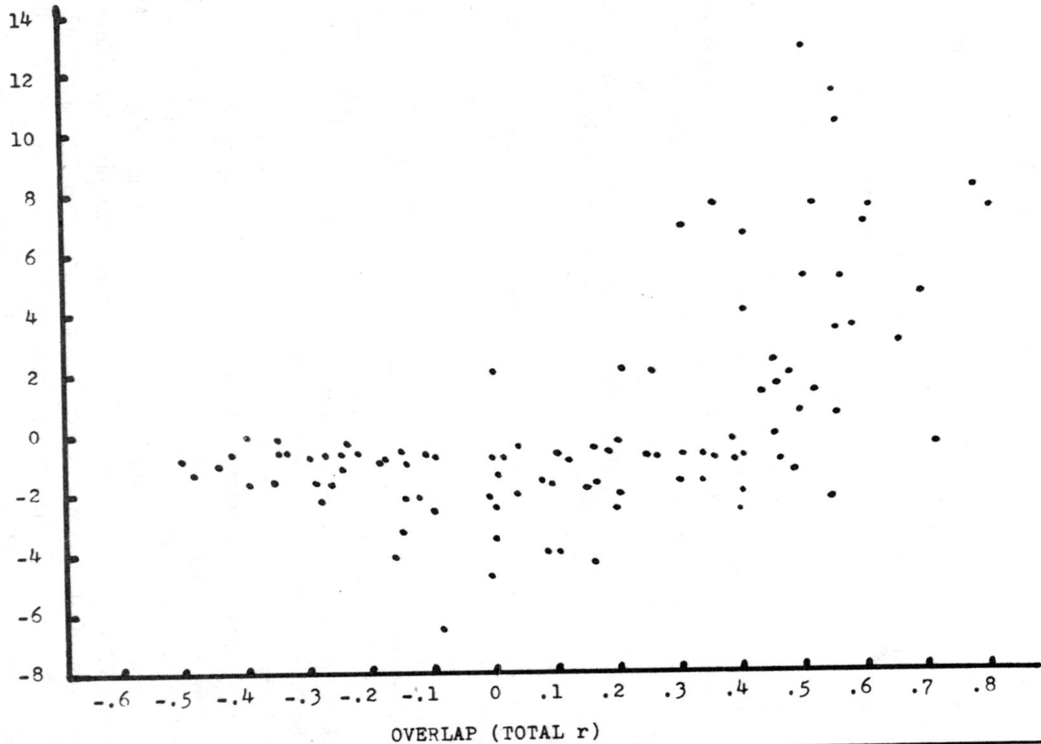

Figure 2: Diplomatic residuals from null model, 100-pair sample

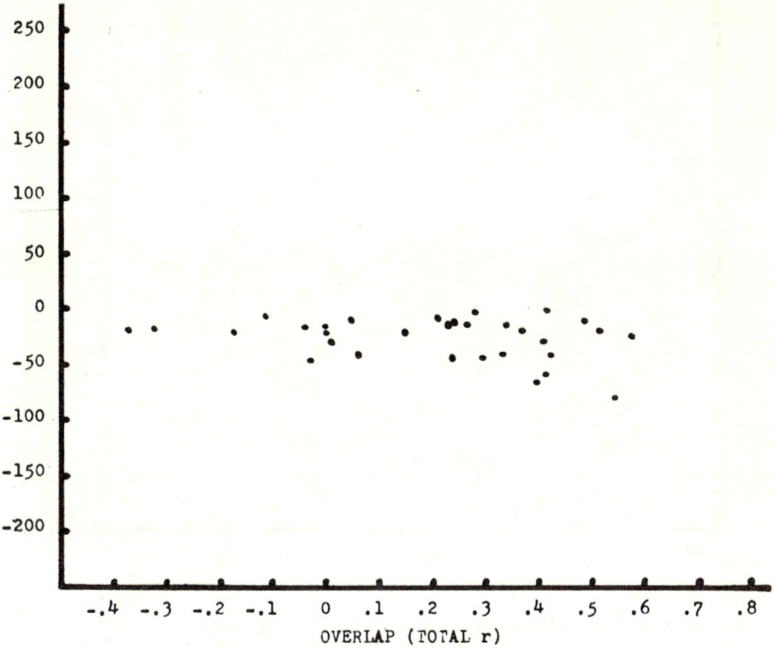

Figure 3a: Trade residuals from null: pairs with link = zero only

view is not tied to a rationale stressing differences between interacting and noninteracting pairs, and the strength of the atomistic prediction does not increase if only nonzero pairs are considered—in fact the correlations drop a little, probably because the range of variation is somewhat restricted. The systemic hypotheses are tied to such a rationale distinguishing interacting from noninteracting pairs, and their contribution does improve when only interacting pairs are considered.

No comparison of interacting pairs with others will be made in the full data sets, to which we now turn, since this would double the cost of calculation without adding anything to the analysis. The results given here are only a partial check on the Simon-Homans rationale, but they are also a very positive check.

IV. RESULTS FOR THE FULL DATA SETS

In this chapter the final forms of the systemic hypotheses are applied to all the suitable trade and diplomatic data. For a given pair of nations, the

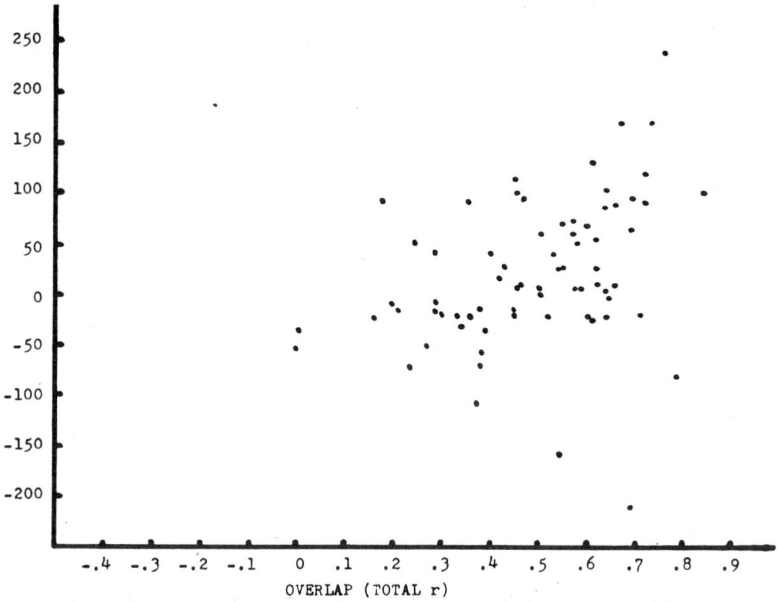

Figure 3b: Trade residuals from null: pairs with link ≠ zero only

observed link (of trade or of diplomacy) is the thing to be predicted. One part of the systemic prediction is overlap, or the rank correlation between the pair members' links with other nations (all those other nations with whom one or both of the pair members have a nonzero link). Another systemic variable is internetwork knitness or the level of interaction between those dyads of nations which (1) are both strongly linked to the focal pair, or (2) are strongly linked as a dyad—that is, one is strongly linked to one pair member, one to the other pair member. In particular "strongly tied" will mean "on the list of top 15 interaction partners" and "level of interaction" will be a resistant measure of level, that is, the trimean. The final systemic variable is intranetwork knitness, or the level of interaction among those dyads of nations strongly tied to one and only one member of the pair; that is the level of links between nations on the first pair member's top list but not the second's, plus the level between nations on the second pair member's list but not the first. As has been argued, overlap (O) represents the strength of the shortest indirect paths between the pair whose link is predicted; internetwork knitness (KI) represents the strength of somewhat longer paths; and intranetwork knitness (KA + KB) represents the strength of clique-like involvements

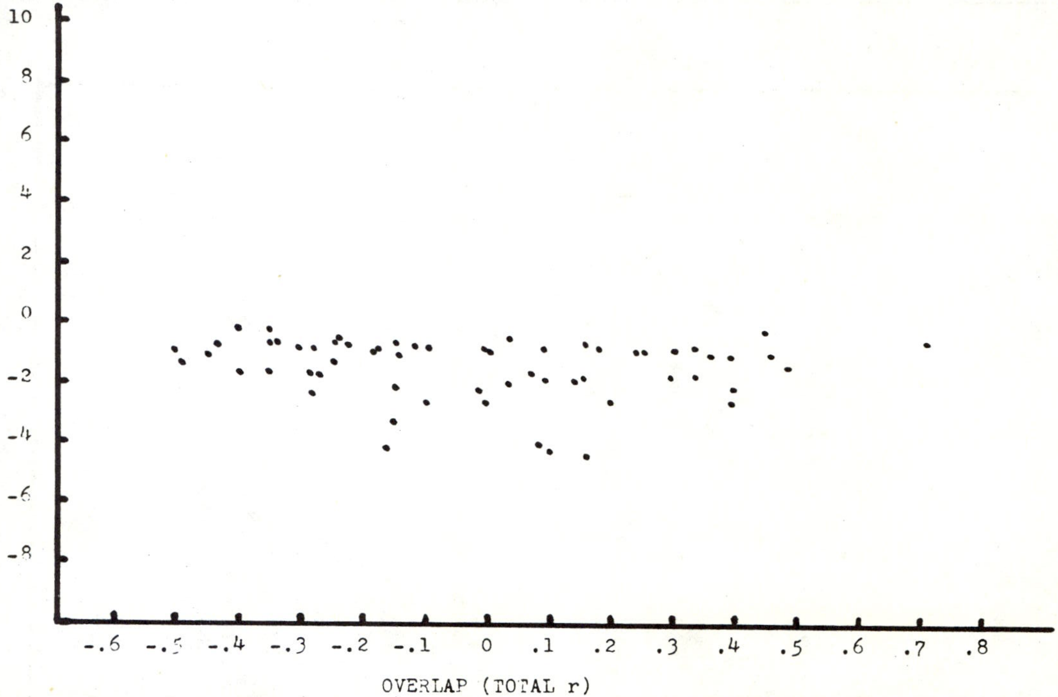

Figure 4a: Diplomatic residuals for sample pairs with zero links

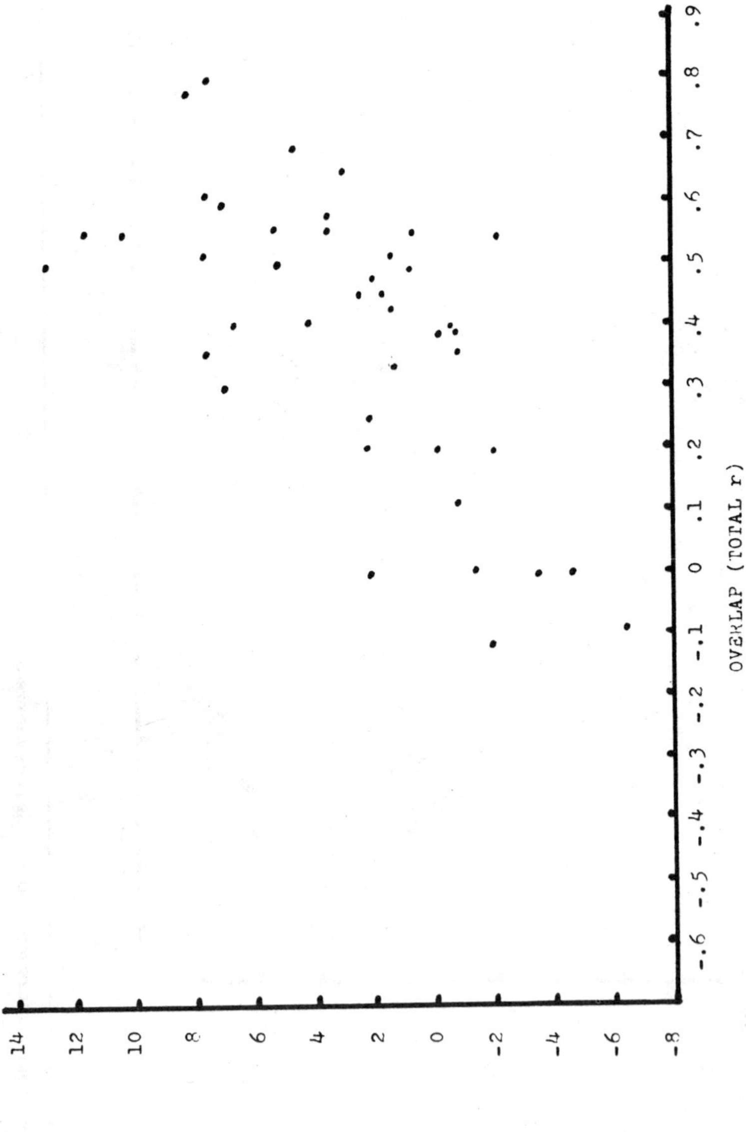

Figure 4b: Diplomatic residuals for sample pairs with nonzero links

TABLE 3
The Null Model With and Without Network Variables

	Trade: All Pairs	Trade: Nonzero	Diplomacy: All Pairs	Diplomacy: Nonzero
R for the full model	.866	.856	.934	.942
r for the null model	.809	.758	.909	.895

"R for the full model" is the multiple correlation with expectations, overlap, KI, KA, and KB as independent variables. The knitness variables are based on a list length of fifteen and on links rather than residuals. KI is based on A_i - B_i as well as C_i - C_i links.

pulling pair members apart. Thus we would expect O to be related positively to pair link size, and KI to be related positively but less strongly (since it represents weaker chains of communication). We expect KA + KB to be related negatively, but we have no obvious analytic reason to guess the strength of its effects (though the work in the preliminary samples suggests a weak effect at best).

First some basic characteristics of the major variables will be reported. Then the important bivariate relationships, like the correlation between overlap and the dependent variable link size, will be given. The systemic variables will be combined in a regression and the more striking deviant cases will be discussed to suggest other important variables which may be operating. Finally, the atomistic variable will be added to the regression in order to get an idea of the relative strength of systemic variables and a non-systemic variable which predicts the strength of a pair's link solely in terms of the overall interaction sizes of the pair members.

The two dependent variables are the nonzero trade and nonzero diplomacy relationships in 1948.

Thus, only about one third of the diplomatic links for the late 1940s are considered, since there were only 998 nonzero links among a total of 3,003 among the 78 nations for which data were available. A rather higher proportion of the 98 trading units had some reported interaction: 2,373 pairs of a possible 4,753. It turns out that the main variables are not always well-matched in form. Discrepancies in distributions can be important in analyses that use correlations heavily, since the possible values of the correlations can be limited by shape differences among the variables. The major features of the distributions of the main variables are summarized in Table 2. Some features are common to both data sets. Both null models tend to underestimate nonzero links because both make a number of small nonzero predictions for pairs of nations which do not in fact interact at all, thus reducing the possible size of predictions for nation pairs which do interact. Both sets of expectations reproduce the spread of the links they model quite accurately, but fail to model higher moments well. Both are more skewed to the right about their means (inclined to have a straggling trail of high values), and both are more strongly peaked than links are. All the same, the overall correspondence of the null models and the links is very good, especially for the diplomatic data. Trade expectations are too low for the trade links on the whole, but also include some values that are far too high for any of the links. There is a trade expectation of 1,106! [11]

Unlike the atomistic variable expectation, the two major systemic variables are not well matched to the dependent variable in form. Instead of having an upward skew, overlap and knitness are quite symmetrical. (The overlap values were multiplied through to avoid decimals and thus satisfy the requirements of the program used, so the summary values reported do not equal the original correlational values. For example, the overlap mean was .365 for diplomacy.) The sheer mismatch of form between the links on the one hand, and overlap or knitness on the other, might obscure part of relationships between them.

Bivariate Relationships

Just how do the links relate to the rival explanations, the atomistic and the structural? The correlation of expectations with links is .90 for diplomacy and .70 for trade. Perhaps the poorer match of distributions in the trade case did make a difference. In any case, the null model accounts for a lot of the link variance in both data sets. Overlap makes a more modest but still notable impact, correlating at .36 with diplomatic links and at .58 with trade links. The KI knitness variable has corresponding correlations of .08 and .23, while the disjoint knitness KA + KB has –.09 and –.01. All of the overlap and knitness relationships to links are in the direction predicted by the main hypotheses, but they tend to be disappointingly weak for the diplomacy data and the knitness variables. So far, it seems that diplomatic ties are relatively more responsive to the mechanisms behind the atomistic hypothesis, and trade ties are more responsive to those behind the structural hypotheses. Perhaps decisions about the sizes of diplomatic missions are strongly affected by the joint prestige of the two nations concerned, where prestige is reflected in a nation's overall rate of diplomatic interaction. For example, there may be some strong hierarchical norms governing the appropriate level of missions between two nations of given levels of importance in foreign affairs (see

Singer and Small, 1966). On the other hand, trade decisions tend to be made by people who want to make money and who respond to information and other pressures thought to make the structural hypotheses work. In diplomacy, the rules of the game (such as protocol) may affect links directly, while the rules of the trading game often stress the overall trade balance rather than the particular link sizes on which it is based.

Combining Systemic Variables

When overlap and KI are used together, they still account for more trade than diplomatic variance: 36% versus 22% (the multiple Rs are .60 and .47). KI adds more to overlap in the diplomatic case, perhaps because there is so much more room for improvement. KI's somewhat disproportionate effects can be seen best from the standardized regression equations:

$$\text{predicted trade} = .71\ (\text{Overlap}) - .20\ (\text{KI}) + 35$$

$$\text{predicted dipl.} = .69\ (\text{Overlap}) - .43\ (\text{KI}) + 5$$

As in the preliminary samples, the weights for KI are negative when overlap is controlled for and links are the dependent variable. This result does not endanger the hypothesis predicting that links increase as KI does. They do increase with KI, as the zero order correlations showed. However, the hypothesis might have to be modified to cope with the facts that (1) zero order correlations are weak, and (2) knitness has one direction of effect without overlap controls and another direction with such control. As for KA + KB, that aspect of knitness which is hypothesized to pull nations apart, results are even weaker. Adding KA + KB to KI and overlap does not change the multiple correlation (the amount explained is not increased) and the beta weights for intranetwork knitness are negligible (–.03 for both trade and diplomacy). The standardized regression equations are:

$$\text{predicted trade} = .71\ (0) - .20\ (\text{KI}) - .03\ (\text{KA+KB}) + 47$$

$$\text{predicted dipl.} = .68\ (0) - .43\ (\text{KI}) - .03\ (\text{KA+KB}) + 6$$

showing that even the beta weights for overlap and internetwork knitness are not affected by the addition of intranetwork knitness.

So far, it is clear that overlap and knitness together account for respectable slices of the link variance, 22% for diplomacy and 36% for trade. But it is also clear that the rival model accounts for much more:

79% for diplomacy and 49% for trade. The structural model has some use, especially for trade–why not more? One must consider the possibility of a curvilinear relationship between the structural model and links, since the linear regressions used would miss or underestimate such relationships. If curvilinearity is a problem here, it could be corrected easily with further transformations of the data. However, examination of plots and printout showed no signs of curvilinear relationships. The examinations did underline a different problem: high scores on the independent variables are associated with a wide range of scores on the dependent variable, especially for diplomacy. The larger the link size which is predicted, the more spread out the observed values and the worse the fit.

Just which pairs are involved where the fit is poor? Table 5 lists the 35 pairs which most exceeded the diplomatic link size predicted (from the United States and United Kingdom with 136 diplomats more than predicted, to the United Kingdom and the Netherlands with 22 more); it also lists the pairs which fell farthest below prediction (from Poland and Luxembourg with 20 too few to the Netherlands and Denmark with 12 less than predicted). Table 6 presents comparably deviant cases for trade, giving a few more, since the trade pairs are so much more numerous.

Overall, the nation pairs with more trade or diplomacy than predicted almost always include a great power. Most of them for trade, and all of them for diplomacy, are pairs consisting of a great power and one of its allies or a great power and another great power. Now the great powers are precisely the nations who have the resources and the desire to shape interaction links in line with current policies. Thus, it was quite feasible for the United Kingdom to strengthen diplomatic ties with Commonwealth nations, even though this involved shifting dozens of diplomats from "predicted" assignments. A smaller nation would not have had enough diplomats to be able to shift so many. To be sure, some of the manpower is Canadian and so forth, but the main impetus was probably British. Otherwise, why do not Commonwealth pairs other than those including the UK appear as high deviates? In trade, the pattern of political pressures is more complex, but still visible: the U.S.S.R. tends to favor East Europe; the USA, Latin America.

The low deviate cases tell a very different story, or rather they differ by telling no story at all. While the high deviant cases look quite similar for trade and for diplomacy, the low cases do not; not a single pair appears in both "smaller than predicted" columns in Table 5 and 6. No particular kind of pair appears with striking frequency; some of the pairs consist of nations from the same region (like two Latin American nations), some do

TABLE 4
Summary Statistics for the Main Variables

DIPLOMACY

Variable	Mean	Standard Deviation	Skewness	Kurtosis
Links	9.58	12.40	8,120.67	793,670.00
Expectation	7.25	12.41	13,622.47	2,021,627.00
Overlap	36,492.70	24,013.41	7,626,112.00	3,412,608.18
Knitness (KI)	30.71	18.82	2,557.38	252,645.00

TRADE

Variable	Mean	Standard Deviation	Skewness	Kurtosis
Links	124.3	96.4	452,358	1,388,160
Expectation	101.4	98.5	2,765,319	2,897,280
Overlap	43,668.4	18,009.8	5,207,680	7,060,095
Knitness (KI)	770.1	201.5	1,184,141	2,250,496

not; nations from all areas of the world are represented, and so on. It is true that the great powers appear very rarely, which suggests that they do not divert trade or diplomacy away from nations as plainly as they divert trade or diplomacy towards certain favored partners.

Finally, there may be some awkward errors in some of the trade link sizes for dependent countries. Their reported link sizes tend to be extreme, either larger than either the systemic or atomistic prediction (like five cases including the Antilles) or smaller than either (Benelux-Trinidad, Norway-Cypress). It is all too likely that trade reports for such areas are highly abbreviated, giving careful data for major trading partners like the colonial country and poorer data for others. Often trade with a dependent area is not reported at all at either end of the transaction. If only the main trade flows get into the statistics, one would expect that major statistics would be underpredicted and minor ones overpredicted, as seems to be the case.

Combining the Atomistic and Systemic Variables

The final step in the full data analysis is the addition of the atomistic variable to the regressions. For the nonzero trade pairs, the beta weights are

$$\text{predicted trade} = .56(\text{Exp}) + .37(\text{O}) - .02(\text{KI}) - .09(\text{KA+KB})$$

The most important single part of the combined prediction is clearly the atomistic part, the expected size of the link if only the interaction sizes of

TABLE 5
Deviant Cases for Systemic Variables Versus Diplomatic Links

Smaller than Predicted	Larger than Predicted
Luxembourg - Poland	USA - UK
Iraq - Saudi Arabia	USA - France
Iceland - Sweden	USA - Italy
Colombia - Haiti	UK - France
Libya - Spain	USA - USSR
Luxembourg - USSR	USA - Sweden
Italy - Rumania	USA - Mexico
Iceland - Poland	USA - Argentina
Argentina - Haiti	USA - Norway
Libya - Netherlands	USA - Cuba
El Salvador - Honduras	USA - Netherlands
Chile - Costa Rica	USA - Spain
Belgium - Libya	UK - Canada
Iceland - USSR	UK - USSR
Argentina - Ireland	USA - Brazil
Italy - Sweden	USA - Belgium
Belgium - Philippines	USA - Taiwan
Finland - Netherlands	USA - Philippines
Bulgaria - Netherlands	USA - Egypt
France - Philippines	USA - Greece
Argentina - Colombia	USA - Canada
Belgium - Ethiopia	UK - Spain
Bolivia - Italy	UK - Italy
Rumania - USSR	USA - Turkey
New Zealand - USSR	UK - Greece
Luxembourg - Switzerland	UK - Australia
Costa Rica - El Salvador	UK - Argentina
Chile - Dominican Republic	USA - Chile
Lebanon - Netherlands	UK - Pakistan
Italy - Portugal	UK - Thailand
Jordan - Saudi Arabia	USA - Iran
Nepal - USA	USA - Venezuela
Albania - Bulgaria	USA - Yugoslavia
Denmark - Netherlands	UK - Netherlands

the pair members are considered. But it is also clear that overlap makes a substantial contribution as well. The weak showing of the two knitness variables, which is not surprising by now, may reflect either their relatively poor measurement or the relative weakness of the more indirect paths which they sum up.

For the nonzero diplomacy pairs,

predicted dipl. = .85(Exp) + .20(O) – .05(KI) – .03(KA+KB)

TABLE 6
Deviant Cases for Systemic Variables Versus Trade Links

Smaller than Predicted	Larger than Predicted
Venezuela - Costa Rica	USA - S Africa
Tanganyika - S Rhodesia	USA - Australia
Honduras - Nicaragua	Antilles - Venezuela
Portugal - Paraguay	USA - Canada
Tanganyika - N Rhodesia	USSR - Bulgaria
Norway - Algeria	Antilles - UK
Italy - Jamaica	Cuba - Japan
Costa Rica - Ecuador	USSR - Saudi Arabia
Bolivia - Uruguay	USA - UK
Greece - Spain	UK - Cuba
Guatemala - Nicaragua	UK - Australia
Paraguay - Spain	UK - Nigeria
Chile - Paraguay	USA - France
Greece - Hungary	USA - India
France - Jordan	USA - Brazil
Guiana - Surinam	USSR - Finland
Pakistan - New Zealand	USA - Germany (FR)
Cuba - Haiti	UK - Canada
Spain - Algeria	Antilles - Argentina
Norway - Sudan	Antilles - Brazil
Paraguay - Norway	UK - Argentina
Spain - Angola	USSR - Poland
Benelux - Trinidad	USA - Venezuela
Costa Rica - Cuba	USSR - Japan
Norway - Cyprus	Germany (FR) - Cuba
Turkey - Jordan	USSR - Mexico
Cuba - Dominican Republic	USA - Congo
Switzerland - N Rhodesia	UK - France
Spain - Ceylon	Benelux - Congo
Norway - Tanganyika	USA - Malaya
Panama - Venezuela	USA - Argentina
Cuba - Paraguay	USA - Colombia
S Africa - Pakistan	USSR - Malaya
Japan - Sarawak	USA - USSR
Jamaica - Sweden	USA - Ghana
Sweden - Tanganyika	France - Canada
Argentina - Mozambique	Canada - Australia
Sweden - Guiana	Antilles - Nigeria
Costa Rica - Argentina	UK - Venezuela
Switzerland - Jamaica	Chile - Egypt
Switzerland - Jordan	Argentina - Japan
Pakistan - Indonesia	USA - Cuba
Cuba - Guatemala	USA - Antilles
Jamaica - Germany (FR)	USA - Greece
Switzerland - Burma	Cuba - India
	France - Argentina

Once again it seems that diplomacy is more responsive to the atomistic and less to the systemic, compared to trade; the importance of overlap is much less. However, overlap does make some impact even though a modest one.

The difference in trade and diplomacy also shows up in the aggregate. For trade, the atomistic variable alone accounts for 49% of link variance (r = .70), while the atomistic and systemic variables together account for 60%, considerably more (multiple R for the regression above = .78). But for diplomacy, the atomistic variable alone accounts for 81% of the link variance (r = .90) while the atomistic and systemic variables together account for a virtually identical 83% (multiple R for the regression above = .91). The systemic variables do have effects in both diplomacy and trade, as the weights for overlap indicate, but the effect is stronger for trade however the effect is evaluated.

Trade and diplomatic relationships are affected both by individual attributes of the nations concerned (atomistic variables like interaction sizes) and by systemic attributes of the web of relationships surrounding each pair of nations. The relative importance of atomistic and systemic variables depends on the kind of relationship examined. Some nations, the handful with greatest power, seem freer than others to escape the constraints of size and system; these nations are the ones most likely to have relationships much different from what one would expect given system or size or both.[12]

V. DISCUSSION

The strength of the link between two nations is a function of three broad classes of variables: classes which vary in the amount of variance they can explain and in the conditions under which they apply. Links depend in the first place on the overall interaction sizes of the two nations concerned, whatever the two nations may be. The link also depends on how the two nations are located in the web of links among others, a relationship that is limited to pairs that do in fact interact to some extent. Finally, link size can vary in response to deliberate manipulation by nations powerful enough to employ these tactics, a factor which applies only to nation pairs including at least one great power.

Each of these variables or variable sets represents a different process, even though the processes impinge on each other enough to produce relationships among the variables. Clarification of these processes is one of the many next steps that need to be taken. So far, it appears that the effects of overall interaction sizes are both the most obvious and the least understood. Most studies control for the activity levels of the members of a pair in one way or another, but few are really helpful in explaining why this should work.

At the beginning of the study, it was enough to know that it should, since this part of the analysis was to be only an atomistic rival for the systemic hypothesis. But now it seems more reasonable to use the atomistic variable as a partner rather than a rival in the explanation of link sizes. Just how does the atomistic variable based on interaction sizes have its effects?

Two somewhat different suggestions have arisen in studies of trade and diplomacy. Trade analysts like Linneman suggest that overall trading level, or the closely related variable of domestic market size, sums up the trade "pull" of a nation–the number of attractive opportunities, as well as its "push"–the number of ready supplies for such opportunities in another country. Combine the economic strengths of two nations and you should have a good estimate of the amount of trade between them. In short, the expected value of a link is an index of the overall resultant of many microprocesses: many buyers looking for many sellers. If this is the underlying process, it could be studied further with more detailed models of how it works, or at the very least, with less crude indices. How many import-export firms are there? How much of the governmental structure revolves around trade? How much of the economy is geared to export or dependent on import?

On the other hand, the connection between expected link sizes and diplomatic links is more likely to be a macroprocess. Decision-makers in the diplomatic service, or the appropriate ministry of a country, rank the other nations they recognize in order of diplomatic importance and send an appropriate share of the available diplomatic staff to each, which explains our results if diplomatic importance is well-indexed by the overall size of a nation's diplomatic service. There is no cumulation of a number of little decisions, but a pair of literally global ones made by the parties in a nation dyad. The point of the decisions is not to make profit for many subsystems, but to make the most effective use of a society's resources, given the diplomatic stratification system of the world. One must assume rational diplomatic decision-makers for this process.

Probably both macro and micro processes are involved in both trade and diplomacy. Trade is, in part, affected by global decisions (which is why Linneman includes "trade preference agreements" as a variable in his model); diplomacy is in part affected by resultants of microprocesses (if only because it is affected by internation links built on such processes–links like trade itself). But it is also probable that the two processes are of unequal importance for the two sets of relationships. The macroprocess, involving a global decision to maximize nation interest, is likely to have a stronger effect on diplomacy than on trade because:

(1) diplomatic link sizes can be set by some governmental agent, while trade sizes rarely can be more than modified in this way;

(2) the criteria for setting the link sizes are clearly spelled out for diplomacy, in which the relative ranks of nations and the importance of tying diplomatic attention to rank, are both very explicit [Singer and Small, 1966].

Now the macroprocess is a much simpler and more direct link between interaction sizes and pair links than the microprocess, so one would expect the atomistic variable "expected link" to be more strongly related to pair links when the macroprocess was operating than when the microprocess was. And it has already been noted that the atomistic variable accounts for more link variance for diplomacy (80% of the variance in nonzero links) than for trade (50% of the variance in nonzero links).

Finally, note that the mechanisms tentatively suggested for atomistic size effects do not seem to have much to do with whether pairs have nonzero interactions or not; indeed the atomistic variable relates well to nonzero and zero links. There is a little puzzle here: if two noninteracting nations grow in size till it is time (according to the atomistic prediction) for them to start a nonzero link, how can it get started? How do those who make the appropriate decisions learn that the time has come, for example? This is especially puzzling for microprocess decisions, involving more complex and less readily available information. One solution to this puzzle is systemic: in the absence of direct information lines through a pre-existing relationship, news passes indirectly through the surrounding neighborhood or relationships.

The systemic variables formed a second broad class of variables related to link size. They were useful in understanding a more restricted range of links, being applicable only to nonzero links while the atomistic variable was related to all links, and their effect was not as large as that of the atomistic variable; nonetheless the systemic variables made a substantial contribution toward explaining the strength of interaction between nations with nonzero interactions. By far the most important systemic variable proved to be overlap, which indicates the strength of the most direct paths of indirect communication between a pair of nations. In light of the rationale of indirect communication, it is reasonable that overlap be more important than knitness. It is also reasonable that intranetwork knitness be very weak in its negative effects on link size, since this can be explained in terms of the same rationale used to justify working with raw links instead of normed ones. That is, a high level of intranetwork knitness may not pull a pair apart because they may respond by increasing their interaction capacities in order to interact both with the cliques tapped by intranetwork knitness and with each other (if overlap and KI are high). However, the weak performance of internetwork knitness and its failure to add much to overlap is surprising; one would expect it to perform more weakly than overlap but not to be as weak as it is. It is quite possible that the problem is a measurement problem, and more adequate approaches to operationalizing knitness should be tried.

Overall the systemic results were consistent with the rationale based on two components: the concept of indirect communication and the Simon-Homans model of interaction. If the systemic variables do sum up indirect communication paths, then overlap should be stronger than either knitness variable, as it is; if the indirect interaction in these channels works as in the Simon-Homans model, then the systemic variables should work for pairs with nonzero interaction but not for those with zero interaction; this prediction is also confirmed. Further tests of this rationale can be made, for example, by looking at later years to see what becomes of pairs not interacting in 1948. A link remains zero until the systemic variables reach some threshold value which is greater than the amount needed just to maintain a link once begun. Thus noninteracting pairs with higher levels of knitness and overlap should be the first to begin interactions in later years. This is a testable prediction since data for later years exist and quite a few of the zero link pairs become nonzero link pairs in these later data sets.

The null model and the structural model together account for virtually all of the link data. However, there is some remaining variation which appears to be a result of the third broad class of variables: allocation of interaction resources to one's own sphere of influence. This occurs only for nations strong enough to have spheres of influence and great resources. One is reminded of the diplomatic null model mechanism because strategic decisions are made on behalf of a nation; and one is reminded of the network variables mechanism because the spheres of influence consist of groups of nations rather than dyads. The linkages in spheres of influence probably have to be understood as systems of relations, not as collections of bonds based on nation traits like interaction volumes.

One might try to pull the three groups of variables (network, null, and strategic) together by treating them all as influences on decision-making. Then it would be vital to distinguish macro- from micro-level decisions. The third layer of strategic considerations and the diplomatic null are primarily macro-level, while the effects of the trade null are more mixed.

The macro-level components could be removed at least approximately by estimating and deleting the effects of government-level decisions, such as trade preference agreements and buy-back provisions attached to foreign aid. One might also want to do some sort of trade breakdown in order to find out the sum of the mini-expected values between a pair for *compatible* commodities. If two nations have a large joint market, but duplicate rather than supplement each other, their trade may not be large—which helps to account for the fact that the major industrialized nations trade less than one might expect. On the other hand, two smallish nations may have needs and supplies that fit like hand in glove.

Thus a good deal more work could be done on the hypotheses developed here; operationalizations could be improved, mechanisms could be further refined and tested, and other relationships of other time periods could be examined. One could go in a rather different direction by trying out the present approach with one major shift: using some form of normed links instead of raw values. Some tentative and partial excursions of this type were made in doing this study, but no thorough investigation has been made here. One possible technique which uses a kind of link norming, and also stresses something like indirect communication, can be found in Beshers and Laumann (1967). Their approach could be seen as a way to replace the two variables, overlap and knitness, with a single estimate of the overall strength of indirect interaction. One finds all the chains of indirect links between A and B through other nations, weights them by the strength of the links involved and the shortness or immediacy of the chain, and adds the weighted chains together. The result is called a "mean first passage time." This is a reasonable way of combining the indirect communication channels and it makes use of much more information about them than the present systemic hypotheses. One serious problem with the Beshers and Laumann model (or most other preset models) is a reliance on strong untested assumptions about how various elements of indirect chains should be weighted and put together. It is abundantly clear from the present study that overlap and knitness have variable strengths, and the matter needs to be explored further before the structural variables are combined in any fixed way. [13]

In general, we do not yet know enough about the international system to work with fixed complex models of the relationships among international relationships. We need more work on flexible building-block systemic variables which tap important aspects of international systems. In this study, preliminary as it is in many ways, some basic systemic concepts have been developed and have proven to be useful in predicting how strongly nations interact in the trade or the diplomatic system. From work on international systems or from work on interpersonal systems or from both, more systemic hypotheses can be derived; there is no lack of data to test these hypotheses.

NOTES

1. Some interesting variations on this basic theme can be found in Foster et al. (1963), in which the likelihood of one person choosing another as a friend is related to their choice relations with others—that is, the extent to which the network is biased in favor of inbreeding of choices. A similar view is given by Davis and Leinhardt (1971), in which a pattern of positive links within but not between cliques is combined with asymmetric upward positive links between cliques at different levels.

2. For a detailed discussion of debates and developments in this literature see Bott's "Reconsiderations" chapter (Bott, 1971).

3. However, some international scholars take a very different view on this point. For example Brams (1966) argues for the use of "relative acceptance" or the extent to which nations interact more or less than one would expect, in part on the grounds that a nation cannot be attentive to all others.

4. The most familiar statistical tests make inferences about a population from a sample's results; here we study the population itself, so results are simply reported. Tests of association, significance levels, and so on have no role here. However, another stream of statistical work (largely from econometrics) deals with something that *is* relevant: measurement error. Given the varying criteria and frequently low standards of national statistics-gathering plus the further biases of reporting plus simple mistakes, measurement error is a serious problem for international data. Unfortunately the problem is beyond this study's scope and has been largely ignored (as in most analyses).

5. Holland and Leinhardt (1974) have a useful bibliography of interpersonal studies of "local structure," that is, structure at the intermediate level chosen by the present study. The considerable literature in elite and corporate structure might be useful, especially in dealing with international organizations, since the duality of elite persons, elite positions, or of directors' companies is paralleled by the duality of nations: international organizations. See Laumann and Pappi (1973).

6. As Bott recounts in "Reconsiderations" (1971), interpersonal researchers have been able to achieve either a comfortable sample size or valid (hence time-consuming) measures—but not both. Data collection and analysis have been more satisfactory for limited subgroups like elites (consider references in previous footnotes), but even for these the measures of strength of interpersonal relationship have been quite crude.

6A. Again, Brams (1966) offers a good illustration of this line of thought. To find a pair's "relative acceptance," first find their expected interaction size using the null model developed by Savage and Deutsch and modified by Goodman (see Goodman, 1963). This null is somewhat more complicated than the atomistic expectations used here, but has the same basic reliance on marginals or total interaction sizes of a pair of nations. Subtract the null expectation from the observed link to get deviations parallel to this analysis' residuals, and finally get relative acceptances by taking the ratio of deviations to expectations. As Brams argues:

> The predicted (or expected) diplomatic exchanges can serve as a *baseline* against which to compare the actual exchanges in order to determine those cases where there is a greater-than-expected exchange of diplomatic personnel and, presumably, a heightened awareness of a country's decision-makers for the affairs of another country [Brams, 1966: 883].

Note that the plausible arguments for approaches like Brams' have not been tested. True, normed and non-normed versions of the same data have been analyzed, but not in ways comparable either to each other or to the procedures used in this study. Brams and Russett both examined the 1963-1964 equivalent of the diplomatic data used here, but Brams hierarchically decomposed saliences (relative acceptances above a set level) while Russett applied direct factor analysis to links (more precisely, their square roots). Comparison of the two sets of results is difficult, since somewhat different questions were asked; in any case, differences could arise from several sources: the norming or non-norming of the links, the use of saliences versus the use of more finely graded data, the use of one type of analysis rather than another.

7. Some readers have criticized this decision on the grounds that extreme values can be the most important ones and a product-moment correlation would very properly reflect the impact of such values. For example, it is not unusual to find a pair of Latin American nations with very little in common but an enormous interaction with the United States–a classic "one point correlation" if product-moment correlation is used. In such cases the one point correlation may capture an important thing (mutual dependence on the same great power) which a rank-order correlation would miss. However, one has no reason to believe that all or even most of the extreme or flukey values will be this meaningful.

8. Many have remarked that there probably is such a neighborhood beyond which it is pointless to gather further records of links (Bott, 1971: 292); one might adapt Epstein's (1969) terms "effective" and "extended" to refer to that neighborhood of links which count and the further reaches of relationships which do not. No one suggests just how to decide where the effective network leaves off.

9. Tukey (1970) suggests that one of the best compromises between maximum use of data information and maximum resistance to extreme values is the trimean:

$$\text{Trimean} = \frac{2\text{Md} + Q_1 + Q_3}{4}$$

where Md = the median, while Q_1 and Q_3 are the other two quartiles. That is, the average level of the third party interaction is found in a way that trims off the top and bottom quarters of the observations. If the data are well-behaved, the trimean and the ordinary mean will be very close; if the data are highly skewed or have extreme values (not unlikely here), then the trimean will be much closer to the main body of the data, giving a value more typical of more of the links whose level is summarized.

10. The support for the rationale is all the stronger because the difference between zero and nonzero pairs appears to be a real one; that is, it does not seem reasonable to argue that the difference is some sort of artifact. I could find no plausible traces of artifacts like curvilinearity or floor effects, for example.

11. Recall that the figures used throughout the trade analyses are two-decimal logs of hundreds of thousands of dollars, with decimals omitted, so that a large figure in the analyses represents a very large amount of money indeed. For example, 1,106 stands for the log 11.06, while the log of the largest actual trade link is 4.54; translating back into dollars yields a top expected link nearly 10^7 times bigger than

the top observed link. At least some of the oddities in the trade distributions stem from the log transformation. In particular, the smaller links cluster at values that look rather far apart in logs, but were smoothly contiguous in dollars. Going from one to two to three hundred thousand dollars seems smooth enough–and is as detailed as the source permits; the corresponding transformed link values go from one to 30 to 48.

12. Tables 5 and 6 show that the great powers are most often the high deviant cases from the systemic variables. Other similar tables of deviant cases from the regression of atomistic residuals against systemic variables also show that the great powers are the most deviant. In these tables, not shown because they are so repetitive of Tables 5 and 6, high deviant cases are most often great powers and a client, next most often two lesser nations from the same sphere of influence. Low deviant cases tend to be great powers and the clients of different great powers, or pairs of great powers. The deviations from this regression come the closest to fitting Galtung's (1966) predictions; great powers tend to interact more than one would expect with their own clients and less than one would expect with clients of other powers. However, the pattern is mixed–for example, Galtung predicts very high interaction among the great powers themselves, but pairs of great powers tend to interact less than expected.

13. Another difficulty with the mean first passage times approach is that it does not seem to allot a role to the expected size of a link, a matter that we have found to be important. Perhaps this difficulty could be side-stepped by substituting an expected value for the actual link value, which mean first passage times include–after all, that actual link value has to be removed somehow, since it is the thing one wants to predict. Otherwise, one has a "boot-strap" problem: the link to be explained is the most heavily weighted component in the independent variable and any relationship may well be spurious. This contamination of variables is not a problem for Beshers and Laumann, who use mean first passage times as an overall measure of social distance to be related to other distances based on other data. They do not try to predict links.

REFERENCES

ALGER, C. F. and S. J. BRAMS (1967) "Patterns of representation in national capitals and inter-governmental organizations." World Politics 19: 646-663.

ALKER, H. R., Jr. and B. M. RUSSETT (1965) World Politics in the General Assembly. New Haven: Yale Univ. Press.

BALES, R. F., F. L. STRODTBECK, T. M. MILLS, and M. E. ROSEBOROUGH (1951) "Channels of communication in small groups." Amer. Socio. Rev. 16: 461-468.

BESHERS, J. M. and E. O. LAUMANN (1967) "Social distance: a network approach." Amer. Socio. Rev. 32: 225-236.

BOTT, E. E. (1971) Family and Social Network. London: Tavistock.

BRAMS, S. J. (1966) "Transaction flows in the international system." Amer. Polit. Sci. Rev. 60: 880-898.

DAVIS, J. and S. LEINHARDT (1971) "The structure of positive interpersonal relations in small groups," in J. Berger [ed.] Sociological Theories in Progress. Boston: Houghton Mifflin.

EPSTEIN, A. L. (1969) "The network and social organization," in J. Clyde Mitchell [ed.] Social Networks in Urban Situations. Manchester: Manchester Univ. Press.

FOSTER, C. C., A RAPOPORT, and C. J. ORWANT (1963) "A study of a large sociogram II." Behavioral Sci. 8: 56-65.

GALTUNG, J. (1968) "Small group theory and the theory of international relations," in Morton A. Kaplan [ed.] New Approaches to International Relations. New York: St. Martin's Press.

--- (1966) "East-West interaction patterns." J. of Peace Research 2: 146-176.

--- (1964) "A structural theory of aggression." J. of Peace Research 1: 95-119.

GOODMAN, L. (1963) "Statistical methods for the preliminary analysis of transaction flows." Econometrica 31: 197-208.

HAAS, E. B. (1964) Beyond the Nation-State. Stanford: Stanford Univ. Press.

HOFFMAN, S. (1965) The State of War. New York: Praeger.

HOLLAND, P. W. and S. LEINHARDT (1974) "The statistical analysis of local structure in networks," working paper 44. Cambridge: Natl. Bureau of Economic Research.

HOMANS, G. C. (1950) The Human Group. New York: Harcourt, Brace & World.

KAPLAN, M. A. (1968) "International systems." Internatl. Encyclopedia of the Social Sciences 15: 479-486. New York: Macmillan & Free Press.

--- (1961) "Problems of theory building," in Knorr, Klaus, and Verba [eds.] The International System. Princeton: Princeton Univ. Press.

--- (1957) System and Process in International Politics. New York: Wiley.

LAGOS MATUS, G. (1963) International Stratification and Underdeveloped Countries. Chapel Hill: Univ. of North Carolina Press.

LAUMANN, E. O. and F. U. PAPPI (1973) "New directions in the study of community elites." Amer. Socio. Rev. 38: 212-230.

LINNEMAN, H. (1966) An Economic Study of International Trade Flows. Amsterdam: North-Holland.

McCLELLAND, C. A. (1966) Theory and the International System. New York: Macmillan.

MORISETTE, J. O. (1958) "An experimental study of the theory of structural balance." Human Relations II: 239-254.

RUSSETT, B. M. (1967) International Regions and the International System. Chicago: Rand McNally.

SIMON, H. J. (1952) Models of Man. New York: Wiley.

SINGER, J. D. (1974) "System stability and transformation: cybernetic models and operational indicators," paper presented at meeting, Amer. Socio. Assn., Montreal.

--- (1963) "Inter-nation influences: a formal model." Amer. Polit. Sci. Rev. 57: 420-430.

--- and M. SMALL (1966) "The composition and status ordering of the international system, 1815-1940." World Politics 18: 236-282.

TAYLOR, H. F. (1970) Balance in Small Groups. New York: Van Nostrand.

TUKEY, J. W. (1970) Exploratory Data Analysis (prelim. ed.). Reading: Addison-Wesley.

TURNER, C. (1967) "Conjugal roles and social networks: a re-examination of a hypothesis." Human Relations 20: 121-130.

United Nations Statistical Papers (1959) Ser. T, vol. 10, Direction of International Trade. New York: United Nations.

United Nations World (1951) The Diplomatic Yearbook. New York: Funk & Wagnalls.